Number and Numbers

Number and Numbers

Alain Badiou

Translated by Robin Mackay

polity

First published in French as *Le Nombre et les nombres* © Editions du Seuil, 1990.

This English edition © Polity Press, 2008

Polity Press
65 Bridge Street
Cambridge CB2 1UR, UK.

Polity Press
350 Main Street
Malden, MA 02148, USA

ISBN-13: 978-07456-3878-2
ISBN-13: 978-07456-3879-9 (pb)

A catalogue record for this book is available from the British Library.

Typeset in 10.5 on 12 pt Sabon
by SNP Best-set Typesetter Ltd., Hong Kong
Printed and bound in Great Britain by MPG Books Limited, Bodmin, Cornwall

For further information on Polity, visit our website: www.polity.co.uk

This book is supported by the French Ministry for Foreign Affairs, as part of the Burgess programme headed for the French Embassy in London by the Institut Français du Royaume-Uni.
www.frenchbooknews.com

ïi institut français

Contents

Translator's Preface

Alain Badiou's *Number and Numbers*, first published two years after his *Being and Event*, is far from being the specialist work its title might suggest. In fact, it recapitulates and deepens *Being and Event*'s explosion of the pretexts upon which the 'philosophy of mathematics' is reduced to a theoretical ghetto; and their kinship to those reactionary modes of thought that systematically obscure the most pressing questions for contemporary philosophy. Neither does *Number and Numbers* balk at suggesting that even the greatest efforts on the part of number-theorists themselves have fallen short of the properly radical import of the question of number. Badiou's astonishing analyses in the historical section of the book uncover the inextricable bond between philosophical assumptions and mathematical approaches to the problem in these supposedly 'merely technical' works. The aim of *Number and Numbers*, then, is certainly not to mould the unwilling reader into a calculating machine, or a 'philosopher of mathematics': its exhortation is that we (mathematicians, philosophers, subjects under Capital) systematically think number *out* of the technical, procedural containment of which its quotidian tyranny, and the abysmal fear it strikes into the heart of the non-mathematician, are but symptoms. Symptoms, needless to say, whose expression within the situation of philosophy is a pronounced distaste for number-as-philosopheme – whence its recognisable absence in much 'continental philosophy', except where it is pilloried as the very nemesis of the ontological vocation. So if the 'return of the numerical repressed' proposed here will, by definition, excite a symptomatic resistance, for

Badiou it alone can clear the way for the proper task of philosophy; as a working-through of the mathematical ontology presented in *Being and Event*, *Number and Numbers* is a thorough conceptual apprenticeship preparatory to the thinking of the event.

For the great thinkers of number-theory at the end of the nineteenth century, the way to an ontological understanding of number was obscured by calculatory and operational aspects. Today, according to Badiou, the political domination of number under capitalism demands that the project be taken up anew: only if contemporary philosophy rigorously thinks through number can it hope to *cut through* the apparently dense and impenetrable capitalist fabric of numerical relations, to think the event that can 'subtract' the subject from that 'ontic' skein without recourse to an anti-mathematical romanticism.

Whilst this doubtless demands 'one more effort' on the part of the non-mathematician, it would be a peevish student of philosophy who, understanding the stakes and contemplating the conceptual vista opened up, saw this as an unreasonable demand – especially when Badiou offers to those lacking in mathematical knowledge the rare privilege of taking a meticulously navigated conceptual shortcut to the heart of the matter.

Badiou's remarkable book comprises a number of different works – a radical philosophical treatise, a contribution to number-theory, a document in the history of mathematics, a congenial textbook and a subtle and subversive exercise in political theory – whose intricate interdependencies defy any order of priority. The translator's task is to reproduce, with a foreign tongue, that unique voice that can compel us to 'count as one' these disparate figures. In negotiating this challenge, I have sought to prioritise clarity over adherence to any rigid scheme of translation, except where mathematical terminology demands consistent usage, or where an orthodoxy is clearly already in force within extant translations of Badiou's work. In the latter case, my references have been Oliver Feltham's landmark translation of *Being and Event*,[1] with which I have sought to harmonise key terms, Peter Hallward's invaluable *A Subject to Truth*,[2] and Ray Brassier and Alberto Toscano's collection of Badiou's *Theoretical Writings*.[3] Apart from these, in translating chapters 2 and 3 I referred closely to Sam Gillespie and Justin Clemens' translation in UMBR(a), *Science and Truth* (2000). Finally, whilst seeking also to maintain continuity with long-standing English translations of number-theoretical works, some classics in their own right, occasionally the rigour of Badiou's thinking has demanded a re-evaluation of their chosen translations for key terms.[4] Translators also find themselves obliged to arbitrate between a fidelity to Badiou's in many ways admirable indifference to the pedantic apparatus of scholarly citation, and the temptation to pin down the allusions and quotations distributed throughout his work. Badiou's selection of texts is so discerning, however, that it is hardly a chore to return to them. Having thus had frequent recourse to the texts touched on in *Number and Numbers* (particularly in the first, historical part), I have seen no reason not to add citations where appropriate.

One presumes that those self-conscious styles of philosophical writing that necessitate laboured circumlocutions or terminological preciosity on the part of a translator would for Badiou fall under the sign of 'modern sophistry', taken to task herein, as elsewhere in his work. Nevertheless, the aspiration to universal conceptual transparency does not preclude consideration of Badiou as stylist: firstly, as Oliver Feltham has remarked, Badiou's sentences utilise subject/verb order in a characteristic way, and I have retained his tensile syntax whenever doing so does not jeopardise comprehension in translation. Perhaps just as importantly, Badiou does not achieve the deft and good-humoured development of such extremely rich and complex conceptual structures as are found in *Number and Numbers* without a generous and searching labour on behalf of the reader, not to mention a talent for suspense. Although the later sections of *Number and Numbers* may seem daunting, I hope to have reproduced Badiou's confident, meticulous, but never stuffy mode of exposition so as to ease the way as much as possible. In fact, in contrast to his own occasionally chilly edicts, I would venture to suggest that here, 'in his element', Badiou allows himself a certain enthusiasm. One certainly does not accompany him on this odyssey without also developing a taste for the 'bitter joy' of Number.

This translation slowly came to fruition on the basis of a somewhat impulsive decision; it may not have survived to completion without the enthusiasm and aid of an internationally dispersed group of friends and acquaintances, actual and virtual, with whom I shared the work in progress. I would like to extend my thanks to those who helped by pointing out errors and offering advice on the evolving manuscript: Anindya Bhattacharyya, Ray Brassier, Michael Carr, Howard Caygill, Thomas Duzer, Zachary L. Fraser, Peter Hallward, Armelle Menard Seymour, Reza Negarestani, Robin Newton, Nina Power, Manuela Tecusan, Alberto Toscano, Keith Tilford, David Sneek, and Damian Veal. My thanks also to Alain Badiou for his generous help and encouragement, and to the Institution and Staff of the Bodleian, Taylor Institution, and Radcliffe Science Libraries in Oxford. Part of my work on the translation was undertaken whilst in receipt of a studentship from the Centre for Research in Modern European Philosophy at Middlesex University, London.

My greatest debt of gratitude is to Ruth, without whose love and understanding my battles with incomprehension could not even be staged; and to Donald, a great inspiration, for whom the infinite joys of number still lie ahead.

Robin Mackay

0

Number Must Be Thought

0.1. A paradox: we live in the era of number's despotism; thought yields to the law of denumerable multiplicities; and yet (unless perhaps this very default, this failing, is only the obscure obverse of a concept-less submission) we have at our disposal no recent, active idea of what number is. An immense effort has been made on this point, but its labours were essentially over by the beginning of the twentieth century: they are those of Dedekind, Frege, Cantor, and Peano. The factual impact of number only escorts a silence of the concept. How can we grasp today the question posed by Dedekind in his 1888 treatise, *Was sind und was sollen die Zahlen?*[1] We know very well what numbers are for: they serve, strictly speaking, for everything, they provide a norm for All. But we still don't know what they are, or else we repeat what the great thinkers of the late nineteenth century – anticipating, no doubt, the extent of their future jurisdiction – said they were.

0.2. That number must rule, that the imperative must be: 'count!' – who doubts this today? And not in the sense of that maxim which, as Dedekind knew, demands the use of the original Greek when retraced: ἀεὶ ὁ ἄνθρωπος ἀριθμητίζει[2] – because it prescribes, for thought, its singular condition in the matheme. For, under the current empire of number, it is not a question of thought, but of realities.

0.3. Firstly, number governs our conception of the political, with the currency – consensual, though it enfeebles every politics of the thinkable – of suffrage, of opinion polls, of the majority. Every 'political' convocation, whether general or local, in polling-booth or parliament, municipal or international, is settled with a count. And every opinion is gauged by the incessant enumeration of the faithful (even if such an enumeration makes of every fidelity an infidelity). What counts – in the sense of what is valued – is that which is counted. Conversely, everything that can be numbered must be valued. 'Political Science' refines numbers into sub-numbers, compares sequences of numbers, its only object being *shifts in voting patterns* – that is, changes, usually minute, in the tabulation of numbers. Political 'thought' is a numerical exegesis.

0.4. Number governs the quasi-totality of the 'human sciences' (although this euphemism can barely disguise the fact that what is called 'science' here is a technical apparatus whose pragmatic basis is governmental). Statistics invades the entire domain of these disciplines. The bureaucratisation of knowledges is above all an infinite excrescence of numbering.

At the beginning of the twentieth century, sociology unveiled its proper dignity – its audacity, even – in the will to submit the figure of communitarian bonds to number. It sought to extend to the social body and to representation the Galilean processes of literalisation and mathematisation. But ultimately it succumbed to an anarchic development of this enterprise. It is now replete with pitiful enumerations that serve only to validate the obvious or to establish parliamentary opportunities.

History has drawn massively upon statistical technique and is – even, in fact above all, under the auspices of academic Marxism – becoming a diachronic sociology. It has lost that which alone had characterised it, since the Greek and Latin historians, as a discipline of thought: its conscious subordination to the real of politics. Even when it passes through the different phases of reaction to number – economism, sociologism – it does so only to fall into their simple inverse: biography, historicising psychologism.

And medicine itself, apart from its pure and simple reduction to its scientific Other (molecular biology), is a disorderly accumulation of empirical facts, a huge web of blindly tested numerical correlations. These are 'sciences' of men *made into numbers*, to the saturation point of all possible correspondences between these numbers and *other numbers*, whatever they might be.

0.5. Number governs cultural representations. Of course, there is television, viewing figures, advertising. But that's not the most important thing. It is in its very essence that the cultural fabric is woven by number alone. A 'cultural fact' is a numerical fact. And, conversely, whatever produces number can be culturally located; that which has no number will have no name either. Art, which deals with number only in so far as there is a *thinking* of number, is a culturally unpronounceable word.

0.6. Obviously, number governs the economy; and there, without a doubt, we find what Louis Althusser would have called the 'determination in the last instance' of its supremacy. The ideology of modern parliamentary societies, if they have one, is not humanism, law, or the subject. It is number, the countable, countability. Every citizen is expected to be cognisant of foreign trade figures, of the flexibility of the exchange rate, of fluctuations in stock prices. These figures are presented as the real to which other figures refer: governmental figures, votes and opinion polls. Our so-called 'situation' is the intersection of economic numericality and the numericality of opinion. France (or any other nation) can only be represented on the balance-sheet of an import–export business. The only image of a country is this inextricable heap of numbers in which, we are told, its power is vested, and which, we hope, is deemed worthy by those who record its mood.

0.7. Number informs our souls. What is it to exist, if not to give a *favourable account* of oneself? In America, one starts by saying how much one earns, an identification that is at least honest. Our old country is more cunning. But still, you don't have to look far to discover numerical topics that everyone can identify with. No one can present themselves as an individual without stating in what way they count, for whom or for what they are really counted. Our soul has the cold transparency of the figures in which it is resolved.

0.8. Marx: 'the icy water of egotistical calculation'.[3] And how! To the point where the Ego of egoism is but a numerical web, so that the 'egotistical calculation' becomes the cipher of a cipher.

0.9. But we don't know what a number is, so we don't know what we are.

0.10. Must we stop with Frege, Dedekind, Cantor or Peano? Hasn't anything *happened* in the thinking of number? Is there only the exorbitant extent of its social and subjective reign? And what sort of innocent *culpability* can be attributed to these thinkers? To what extent does their idea of number prefigure this anarchic reign? Did they think number, or the future of generalised numericality? Isn't *another idea* of number necessary, in order for us to turn thought back against the despotism of number, in order that the Subject might be subtracted from it? And has mathematics simply stood by silently during the comprehensive social integration of number, over which it formerly had monopoly? This is what I wish to examine.

I

Genealogies: Frege, Dedekind, Peano, Cantor

I

Greek Number and Modern Number

1.1. The Greek thinkers of number related it back to the One, which, as we can still see in Euclid's *Elements*,[1] was considered not to be a number. From the supra-numeric being of the One, unity is derived. And a number is a collection of units, an addition. Underlying this conception is a problematic that stretches from the Eleatics through to the Neoplatonists: that of the procession of the Multiple from the One. Number is the schema of this procession.

1.2. The modern collapse of the Greek thinking of number proceeds from three fundamental causes.

The first is the irruption of the problem of the infinite – ineluctable from the moment when, with differential calculus, we deal with the reality of *series* of numbers which, although we may consider their limit, cannot be assigned any terminus. How can the limit of such a series be thought *as number* through the sole concept of a collection of units? A series tends towards a limit: it is not the addition of its terms or its units. It cannot be thought as a procession of the One.

The second cause is that, if the entire edifice of number is supported by the being of the One, which is itself beyond being, it is impossible to introduce, without some radical subversion, that *other* principle – that ontological stopping point of number – which is zero, or the void. It could be, certainly – and Neoplatonist speculation appeals to such a thesis – that the ineffable and archi-transcendent character of the One can be marked by zero. But then the problem comes back to numerical one: how to *number* unity, if the One that supports it is void? This problem is so complex that, as we shall see, it remains today the key to a modern thinking of number.

The third reason, and the most contemporary one, is the pure and simple dislocation of the idea of a being of the One. We find ourselves under the jurisdiction of an epoch that *obliges* us to hold that being is essentially multiple. Consequently, number cannot proceed from the supposition of a transcendent being of the One.

1.3. The modern thinking of number thus found itself compelled to forge a mathematics subtracted from this supposition. In so doing, it took three different paths:

Frege's approach, and that of Russell (which we will call, for brevity, the logicist approach), seeks to 'extract' number from a pure consideration of the laws of thought itself. Number, according to this point of view, is a universal trait[2] *of the concept*, deducible from absolutely original principles (principles without which thought in general would be impossible).

Peano's and Hilbert's approach (let's call this the formalist approach) construes the numerical field as an operational field, on the basis of certain singular axioms. This time, number occupies no particular position as regards the laws of thought. It is a system of rule-governed operations, specified in Peano's axioms by way of a translucid notational practice, entirely transparent to the material gaze. The space of numerical signs is simply the most 'originary' of mathematics proper (preceded only by purely logical calculations). We might say that here the concept of number is entirely mathematised, in the sense that it is conceived as existing only in the course of its usage: the essence of number is calculation.

The approach of Dedekind and Cantor, and then of Zermelo, von Neumann and Gödel (which we shall call the set-theoretical or 'platonising' approach) determines number as a particular case of the hierarchy of sets. The fulcrum, absolutely antecedent to all construction, is the empty set; and 'at the other end', so to speak, nothing prevents the examination of infinite numbers. The concept of number is thus referred back to an ontology of the pure multiple, whose great Ideas are the classical axioms of set theory. In this context, 'being a number' is a *particular predicate*, the decision to consider as such certain classes of sets (the ordinals, or the cardinals, or the elements of the continuum, etc.) with certain distinctive properties. The essence of number is that it is a pure multiple endowed with certain properties relating to its internal order. Number *is*, before being made available for calculation (operations will be defined 'on' sets of pre-existing numbers). Here we are dealing with an ontologisation of number.

1.4. My own approach will be as follows:

(a)   The logicist perspective must be abandoned for reasons of internal consistency: it cannot satisfy the requirements of thought, and especially of philosophical thought.
(b)   The axiomatic, or operational, thesis is the thesis most 'prone' to the ideological socialisation of number: it circumscribes the question of a thinking of number *as such* within an ultimately *technical* project.
(c)   The set-theoretical thesis is the strongest. Even so, we must draw far more radical consequences than those that have prevailed up to the present. This book tries to follow the thread of these consequences.

1.5. Whence my plan: To examine the theses of Frege, Dedekind and Peano. To establish myself within the set-theoretical conception. To radicalise it. To demonstrate (a most important point) that in the framework of this radicalisation we will rediscover *also* (but not only) 'our' familiar numbers: whole numbers, rational numbers, real numbers, all, finally, thought outside of ordinary operational manipulations, as subspecies of a *unique* concept of number, itself statutorily inscribed within the ontology of the pure multiple.

1.6. Mathematics has already proposed this reinterpretation, as might be expected, but only in a recessive corner of itself, blind to the essence of its own thought: the theory of surreal numbers, invented at the beginning of the 1970s by J. H. Conway (*On Numbers and Games*, 1976),[3] taken up firstly by D. E. Knuth (*Surreal Numbers*, 1974),[4] and then by Harry Gonshor in his canonical book (*An Introduction to the Theory of Surreal Numbers*, 1986).[5] Any interest we might have in the technical details of this theory will be here strictly subordinated to the matter in hand: establishing a thinking of number that, by fixing the latter's status as a form of the thinking of Being, can free us from it sufficiently for an event, always trans-numeric, to summon us, whether this event be political, artistic, scientific or amorous. Limiting the glory of number to the important, but not exclusive, glory of Being, and thereby demonstrating that what proceeds from an event in terms of truth-fidelity can never be, has never been, *counted.*

1.7. None of the modern thinkers of number (I understand by this, I repeat, those who, between Bolzano and Gödel, tried to pin down the idea of number at the juncture of philosophy and the logico-mathematical) have been able to offer a *unified* concept of number. Customarily we speak of 'number' with respect to natural whole numbers,[6] 'relative' (positive and negative) whole numbers, rational numbers (the 'fractions'), real numbers (those that number the linear continuum) and, finally, complex numbers and quaternions. We also speak of number in a more directly set-theoretical sense when designating types of well-orderedness (the ordinals) and pure quantities of any multiple whatsoever, including infinite quantities (the cardinals). We might expect that a concept of number would subsume all of these cases, or at least the more 'classical' among them, that is to say, the whole natural numbers (the most obvious schema of discrete 'stepwise' enumeration) and the real numbers (the schema of the continuum). But this is not at all the case.

1.8. The Greeks clearly reserved the concept of number for whole numbers, which was quite in keeping with their conception of the composition of number on the basis of the One, since only natural whole numbers can be represented as collections of units. To treat of the continuum, they used geometrical denominations, such as the relations between sizes or measurements. So their powerful conception was marked through and through by that division of mathematical disciplines on the basis of whether they treat of one or the other of what were held by the Greeks to be the two possible types of object: numbers (from which arithmetic proceeds) and figures (from which, geometry). This division refers, it seems to me, to the two orientations whose unity is dialectically effectuated by effective, or materialist, thought: the algebraic orientation, which works by composing, connecting, combining elements; and the topological orientation, which works by perceiving proximities, contours and approximations, and whose point of departure is not elementary belongings but inclusion, the part, the subset.[7] This division is still well-founded. Within the discipline of mathematics itself, the two major divisions of Bourbaki's great treatise, once the general onto-logical framework of set theory is set out, deal with 'algebraic struc-tures' and 'topological structures'.[8] And the validity of this arrangement subtends all dialectical thought.

1.9. It is nevertheless clear that, ever since the seventeenth century, it has no longer been possible to place any sufficiently sophisticated mathematical concept exclusively *on one side of the opposition arithmetic/geometry*. The triple challenge of the infinite, of zero and of the termination of the idea of the One disperses the idea of number, shreds it into a refined dialectic of geometry and arithmetic, of the topological and the algebraic. Cartesian analytic geometry radically subverts the distinction from the very outset, and what we know today as 'number-theory' had to call on the most complex resources of 'geometry', in the extremely broad sense in which the latter has been understood in recent decades. Moderns therefore can no longer accept the concept of number as the object whose provenance is foundational (the idea of the One) and whose domain is prescribed (arithmetic). 'Number' is said in many senses. But which of these senses constitutes a concept, allowing something singular to be proposed to thought under this name?

1.10. The response to this question, in the work of the thinkers I have mentioned, is altogether ambiguous and exhibits no unanimity whatsoever. Dedekind, for example, can legitimately be named as the first one to have, with the notion of the cut, convincingly 'generated' the real numbers from the rationals.[9] But when he poses the question: 'What are numbers?' he responds with a general theory of ordinals which certainly, as a particular case, might found the status of whole numbers, but which cannot be applied directly to real numbers.[10] In which case, what gives us the right to say that real numbers are 'numbers'? Similarly, in *The Foundations of Arithmetic*[11] Frege offers a penetrating critique of all previous definitions (including the Greek definition of number as a 'set of units')[12] and proposes a concept of 'cardinal number' that in effect subsumes – on the basis of certain arguable premises, to which I shall later return – cardinals in the set-theoretical sense, of which natural whole numbers represent the finite case. But at the same time he excludes ordinals, to say nothing of rational numbers, real numbers or complex numbers. To use one of his favourite expressions, such numbers do not 'fall under the [Fregean] concept' of number. Finally, it is clear that Peano's axiomatic defines whole numbers and them alone, as a rule-governed operational domain. Real numbers can certainly be defined directly with a special axiomatic (that of a complete, totally ordered Archimedean field). But, if the essence of 'number' resides in the specificity of the statements constituting these axiomatics, then, given that these statements are entirely dissimilar in the case of the axiomatic of whole numbers and of that of real numbers, it would seem that, in respect of their concept, whole numbers and real numbers have nothing in common.

1.11. It is as if, challenged to propose a concept of number that can endure the modern ordeal of the defection of the One, our thinkers reserve the concept for one of its 'incarnations' (ordinal, cardinal, whole, real . . .), without being able to account for the fact that the idea and the word 'number' are used for *all* of these cases. More particularly, they prove incapable of defining any unified approach, any common ground, for discrete numeration (whole numbers), continuous numeration (real numbers) and 'general', or set-theoretical, numeration (ordinals and cardinals). And yet it was precisely the problem of the continuum, the dialectic of the discrete and the continuous, which, saturating and subverting the ancient opposition between arithmetic and geometry, compelled the moderns to rethink the idea of number. In this sense their work, admirable as it is in so many ways, is a failure.

1.12. The anarchy thus engendered (and I cannot take this anarchy to be innocent of the unthinking despotism of number) is so much the greater in so far as the methods put to work in each case are totally disparate:

(a) Natural whole numbers can be determined either by means of a special axiomatic, at whose heart is the principle of recurrence (Peano), or by means of a particular (finite) case of a theory of ordinals, in which the principle of recurrence becomes a theorem (Dedekind).

(b) To engender negative numbers, algebraic manipulations must be introduced that do not bear on the 'being' of number, but on its operational arrangement, on structures (symmetricisation of addition).

(c) These manipulations can be repeated to obtain rational numbers (symmetricisation of multiplication).

(d) Only a fundamental rupture, marked this time by a shift towards the topological, can found the passage to real numbers (consideration of infinite subsets of the set of rationals, cuts or Cauchy sequences).

(e) We return to algebra to construct the field of complex numbers (algebraic closure of the Real Field, adjunction of the 'ideal' element $i = \sqrt{-1}$, or direct operational axiomatisation on pairs of real numbers).

(f) Ordinals are introduced through the consideration of types of order (Cantor), or through the use of the concept of transitivity (von Neumann).

(g)   The cardinals are treated through a totally different procedure, that of biunivocal correspondence.[13]

1.13. This arsenal of procedures was historically deployed according to overlapping lines which passed from the Greeks, the Arab algebraists and those of Renaissance Italy, through all the founders of modern analysis, down to the 'structuralists' of modern algebra and the set-theoretical creations of Dedekind and Cantor. How are we to extract from it a clear and univocal idea of number, whether we think it as a type of being or as an operational concept? All that the thinkers of number have been able to do is to demonstrate the intellectual procedures that lead us to *each* species of 'number'. But, in doing so, they left number as such in the shadow of its name. They remained distant from that 'unique number which cannot be any other',[14] whose stellar insurrection Mallarmé proposed.

1.14. The question, then, is as follows: is there a concept of number capable of subsuming, under a single type of being answering to a uniform procedure, at least natural whole numbers, rational numbers, real numbers and ordinal numbers, whether finite or infinite? And does it even make sense to speak of a number without at once specifying which singular, irreducible apparatus it belongs to? The answer is yes. This is precisely what is made possible by the marginal theory, which I propose to make philosophically central, of 'surreal numbers'.

This theory offers us the true contemporary concept of number, and in doing so it overcomes the impasse of the thinking of number in its modern-classical form, that of Dedekind, Frege and Cantor. On its basis, and as the result of a long labour of thought, we can prevail over the blind despotism of the numerical unthought.

1.15. We must speak not of a single age of the modern thinking of number, but of what one might call, taking up an expression Natacha Michel applies to literature, the 'first modernity' of the thinking of number.[15] The names of this first modernity are not those of Proust and Joyce, but those of Bolzano, Frege, Cantor, Dedekind and Peano. I am attempting the passage to a second modernity.

1.16. I have said that the three challenges to which a modern doctrine of number must address itself are those of the infinite, of zero and of the absence of any grounding by the One. If we compare Frege and Dedekind – so close on so many points – on this matter, we immediately note that *the order* in which they arrange their responses to these challenges differs in an essential respect:

On the infinite   Dedekind, with admirable profundity, *begins with the infinite*, which he determines with a celebrated positive property: 'A system S is said to be infinite when it is similar to a proper part of itself.'[16] And he undertakes immediately to 'prove' that such an infinite system exists. The finite will be determined only subsequently, and it will be the finite that is the negation of the infinite (in which regard Dedekind's numerical dialectic has something of the Hegelian about it).[17] Frege, on the other hand, begins with the finite, with natural whole numbers, of which the infinite will be the 'prolongation' or the recollection in the concept.[18]

On zero   Dedekind abhors the void and its mark, and says so quite explicitly: '[W]e intend here for certain reasons wholly to exclude the empty system which contains no elements at all.'[19] Whereas Frege makes the statement 'zero is a number'[20] the rock of his whole edifice.

On the One   There is no trace of any privileging of the One in Frege (precisely because he starts audaciously with zero). So one – rather than *the One* – comes only in second place, as that which falls under the concept 'identical to zero' (the one and only object that falls under the concept being zero itself, we are entitled to say that the extension of this concept is one). Dedekind, on the other hand, retains the idea that we should 'begin' with one: 'the base-element 1 is called the base-number of the number-series N'.[21] And, correlatively, Dedekind falls back without hesitation on the idea of an absolute All[22] of thought, an idea that could not appear as such in Frege's formalism: 'My own realm of thoughts, i.e. the totality S of all things, which can be objects of my thought, is infinite.'[23] Thus we see that, in retaining the rights of the One, the All is supposed, because the All is that which, necessarily, *proceeds from the One*, once the One is.

1.17. These divergences of order are no mere technical matter. They relate, for each of these thinkers, to the respective centre of gravity of their conception of number and – as we shall see – to the simultaneous stopping point and founding point of their thought: the infinite and existence for Dedekind, zero and the concept for Frege.

1.18. The passage to a second modernity of the thinking of number obliges thought to *return* to zero, to the infinite and to the One. A total dissipation of the One, an ontological decision as to the being of the void and that which marks it, a lavishing without measure of *infinities*: such are the parameters of such a passage. Unbinding from the One delivers us to the unicity of the void and to the dissemination of the infinite.

2

Frege

2.1. Frege[1] maintains that pure thought engenders number. Like Mallarmé, albeit without the effect of Chance, Frege thinks that 'every thought emits a dicethrow'.[2] What is called Frege's 'logicism' runs very deep: number is not a singular form of being, or a particular property of things. It is neither empirical nor transcendent. Nor is it, on the other hand, a constitutive category; it is deduced from the concept. It is, in Frege's own words, a *trait of the concept*.[3]

2.2. The pivotal property that permits the transition from pure concept to number is that of a concept's *extension*. What does this mean? Given any concept whatsoever, an object 'falls' under this concept if it is a 'truth-case' of this concept, if the statement that attributes to this object the property comprised in the concept is a true statement. In other words, if the object satisfies the concept. Note that everything originates with the truth-value of statements, which is their denotation (truth or falsity). It could be said that, if the concept generates number, it does so only in so far as there is truth. Number is in this sense the index of truth, not an index of being.

But the idea of extension is ramified and obscure.

2.3. Given a concept, by the extension of that concept we mean all the truth-cases (all objects qua *truth-cases*) that fall under this concept. Every concept has an extension.

Now, take two concepts C_1 and C_2. We will call them *equinumerate*[4] if there exists a biunivocal correspondence associating, object for object, that which falls under concept C_1 with that which falls under concept C_2. That is, if a biunivocal correspondence can be defined between the extension of C_1 and the extension of C_2.

It is clear that Frege favours a 'cardinal' definition of number; that he is not overly concerned with the structural order of that which falls under the concept. And in fact this essential tool of biunivocity is characteristic of all attempts to 'number' the multiple in itself, the pure multiple subtracted from all structural considerations. To say that two concepts are equinumerate is to say that they have the 'same quantity', that their extensions are the same size, abstracting from any consideration as to what the objects *are* that fall under those concepts.

2.4. Number consists in *marking* equinumeracy, the quantitative identity of concepts. Whence the famous definition: 'The number which belongs to the concept C is the extension of the concept "equinumerate to concept C".'[5] Which means: every concept C generates a number – namely, the set of concepts equinumerate to C, having the 'same pure quantity', the same quantity of extension, as C. Note that a number, grasped in its being, always designates a *set of concepts*, namely all those that satisfy the statement 'is a concept equinumerate to C'.

2.5. The concept of number is constructed through the following progression:

Concept → Truth → Objects that fall under the concept (that satisfy the statement attributing the concept to the object) → Extension of the concept (all truth-cases of the concept) → Equinumeracy of two concepts (via biunivocal correspondence of their extensions) → Concepts that fall under the concept of equinumeracy to a given concept C (that satisfy the statement 'is equinumerate to C') → The extension of equinumeracy-to-C (the set of concepts from the preceding stage) → The number that belongs to concept C (number is thus the name for the extension of equinumeracy-to-C).

From a simplified and operational point of view, it could also be said that, starting from the concept, we are able to pass through the object on condition that there is truth; that we then compare concepts, and that number names a set of concepts that have in common a property made possible and defined by this comparison (equinumeracy).

2.6. To rediscover the 'ordinary', familiar numbers on the basis of this pure conceptualism regulated by truth alone, Frege begins with his admirable deduction of zero: zero is the number belonging to the concept 'not identical to itself'.[6] Since every object is identical to itself, the extension of the concept 'not identical to itself' is empty. It follows that zero is the set of concepts whose extension is empty and which, by virtue of this, are equinumerate to the concept 'not identical to itself'. Which is precisely to say that zero is that number belonging to *every* concept whose extension is empty, is zero.

I have indicated in **1.17** the passage to the number 1: 'One' is the number that belongs to the concept 'identical to zero'.[7] It is interesting to note that Frege emphasises, with regard to 1, that it has no 'intuitive' or empirical privilege, any more than it is a transcendent foundation: 'The definition of 1 does not presuppose, for its objective legitimacy, any matter of observed fact.'[8] There can be no doubt that Frege participates in the great modern process of the destitution of the One.

The engendering of the sequence of numbers beyond 1 poses only technical problems, which are resolved, in passing from n to $n + 1$, by constructing between the extensions of corresponding concepts a correlation such that the 'remainder' is exactly 1 – which has already been defined.

2.7. Thus the deduction of number as a consequence of the concept appears to have been accomplished. More exactly: from the triplet concept/truth/object, and from the single formal operator of biunivocal correspondence, number emerges as an instance of pure thought, or an integrally logical production; thought *must* presuppose itself, in the form of a concept *susceptible* to having truth-cases (and therefore endowed with an extension). In so doing, thought presupposes number.

2.8. Why choose particularly the concept 'not identical to itself' to ground zero? Any concept could be chosen so long as one is sure it has an empty extension, that no thinkable object could have the property it designates. For example 'square circle' – a concept which in fact Frege declares is 'not so black as [it is] painted'.[9] Since we seek an entirely conceptual determination of number, the arbitrary nature of this choice of concept is a little embarrassing. Frege is quite aware of this, since he writes: 'I could have used for the definition of nought any other concept under which no object falls.'[10] But, to obviate his own objection, he invokes Leibniz: the Principle of Identity, which says that every object is identical to itself, has the merit of being 'purely logical'.[11] Purely logical? But we understood that it was a matter of legitimating logico-mathematical categories (specifically, number) on the sole basis of the laws of pure thought. Isn't there a risk of circularity if a logical rule is required right at the outset? Now, equality is one of the logical, or operational, predicates that require grounding (namely, equality between numbers). It might be said, of course, that 'identical to itself' should not be confused with 'equal to itself'. But if 'identity' must here indeed be carefully distinguished from the logical predicate of equality, it is nevertheless equally clear that the statement 'every object is identical to itself' is not a 'purely logical' statement. *It is an onto-logical statement.* And, qua ontological statement, it is immediately disputable: no Hegelian, for example, would admit the universal validity of the principle of identity. For our hypothetical Hegelian, the extension of the concept 'not identical to itself' is anything but empty!

2.9. The purely a priori determination of a concept *certain* to have an empty extension is an impossible task without powerful prior ontological axioms. The impasse that Frege meets here is that of an unchecked doctrine *of the object.* For, from the point of view of the pure concept, what is an 'object' in general, any object whatsoever, taken from the total Universe of objects? And why is the object required to be identical to itself, when the concept is not even required to be non-contradictory in order to be legitimate, as indicated by Frege's positive regard for concepts of the 'square circle' type, which, he stresses, are concepts like any other? Why would the law of the being of objects be more stringent than the law of the being of concepts? Doubtless it would be so *if one were to accept Leibnizian ontology*, for which existent objects obey an *other principle* than do thinkable objects, the Principle of Sufficient Reason. It thus appears that the deduction of number on the basis of the concept is not so much universal, or 'purely logical', as it is Leibnizian.

2.10. To posit as self-evident that the extension of a concept is this or that (for example, that the extension of the concept 'not identical to itself' is empty) is tantamount to supposing that we can move unproblematically from concept to existence, given that the extension of a concept brings into play the 'objects' that fall under this concept. A generalised ontological argument is at work here, and it is this very argument that subtends the deduction of number on the basis of the concept alone: number belongs to the concept *through the mediation of the thinkable objects that fall under the concept.*

2.11. The principal thought-content of Russell's paradox, communicated to Frege in 1903, is its undermining of every pretension to legislate over existence on the basis of the concept alone, and especially over the existence of the extension of concepts. Russell presents a concept (in Frege's sense) – the concept 'to be a set that is not an element of itself' – which is surely a wholly proper concept (more so, truth be told, than 'not identical to itself'), but one nonetheless *whose extension does not exist*. It is actually contradictory to suppose that 'objects' – in this instance, sets – that 'fall under this concept' themselves form a set.[12] And, if they do not form a set, then no biunivocal correspondence whatsoever can be defined for them. So this 'extension' does not sustain equinumeracy, and consequently *no number* belongs to the concept 'set that is not an element of itself'.

The advent, to the concept, of an innumerable ruins Frege's general deduction. And, taking into account the fact that the paradoxical concept in question is quite ordinary (for example, the concept is valid for all the sets customarily used by mathematicians: they are not elements of themselves), we might well suspect that there exist many other concepts to which no number belongs. In fact, it is impossible to predict a priori the extent of the disaster. Even the concept 'not identical to itself' could well turn out not to have any existent extension, which is something entirely different from having an empty extension. Let's add that Russell's paradox is purely logical, that is to say, it is precisely proven: to admit the existence of a set of all those sets that do not belong to themselves undermines deductive language by introducing a *formal* contradiction (the equivalence between a proposition and its negation).

2.12. A sort of 'repair' was proposed by Zermelo.[13] It consists in saying that we can conclude from the concept the existence of its extension *on condition that we operate within an already given existence*. Given a concept C *and a domain of existing objects*, we can say that there exists, *in this existing domain*, the set of objects that fall under this concept – i.e. the extension of the concept. Obviously, this extension is relative to a domain specified in advance and does not exist 'in itself'. This is a major ontological transformation: within this new framework it is not possible to move from concept to existence (and thus to number); we can only move to an existence that is somehow carved out of a pre-given existence. We can 'separate' in a given domain those objects within it that validate the property exposed by the concept. This is why Zermelo's principle, which drastically limits the rights of the concept and of language over existence, is called the Axiom of Separation. And it does indeed seem that accepting this axiom safeguards us against the inconsistency-effects of Russell-type paradoxes.

2.13. Russell's paradox is not paradoxical in the slightest. It is a materialist argument, because it *demonstrates* that multiple–being is anterior to the statements that affect it. It is impossible, says the 'paradox', to accord to language and to the concept the right of unfettered legislation over existence. Even supposing that there is a transcendental function of language, it supposes also the availability of some prior existent, the power of this function being simply that of carving out or delimiting extensions of the concept within this specified existent.

2.14. Can we, in assuming Zermelo's axiom, save the Fregean construction of number? Once again, everything turns on the question of zero. I might proceed in the following way: given a delimited domain of objects, whose existence is somehow externally guaranteed, I will call 'zero' (or 'empty set', which is the same thing) that which detaches, or separates, within this domain, the concept 'not identical to itself', or any other such concept under which I can assure myself that no objects of the domain fall. As we are dealing with a limited domain, and not, as in Frege's construction, with 'all objects' (a formulation that led to the impasse of a Leibnizian choice without criteria), there is a chance of my finding such a concept. If, for example, I take a set of black objects, I will call 'zero' that which separates in this set the concept 'to be white'. The rest of the construction will follow.

2.15. But what domain of objects could I start with, for which it can be guaranteed that these objects pertain to pure thought, that they are 'purely logical'? Recall that Frege intends to construct a concept of number that is, according to his own expression, 'not . . . either anything sensible or a property of an external thing',[14] and that he emphasises on several occasions that number is subtracted from the representable. Establishing that number is a production of thought, deducing it from the abstract attributes of the concept in general – this cannot be achieved using black and white objects. The question then becomes: what existent can I assure myself of, outside of any experience? Is the axiom 'something exists' an axiom of pure thought and, supposing that it is, can I discern any property of which it is *certain* that it does not belong in any way to this existent 'something'?

2.16. A 'purely logical' demonstration of existence, for thought, of a nondescript object, a point of being, an 'object $= x$': the statement 'every x is equal to itself' is an axiom of logic with equality. Now, the universal rules of first-order logic, a logic valid for every domain of objects, allow us to deduce, from the statement 'every x is equal to x', the statement 'there exists an x that is equal to x' (subordination of the existential quantifier to the universal quantifier).[15] Therefore, there exists x (at least that x which is equal to itself).

Thus we can demonstrate within the framework of set theory, *first of all*, by purely logical means, that a set exists. *Then* we can separate the empty set within that existent whose existence has been proved, by utilising a property that no element can satisfy (for example, 'is not equal to itself'). We have respected Zermelo's axiom, since we have operated within a prior existent, but we have succeeded in engendering zero.

2.17. It is quite obvious, I think, that this 'proof' is an unconvincing artifice, a logical sleight of hand. From the universal postulate of self-equality (which we might possibly accept as an abstract law, or a law of the concept), who could reasonably infer that there *exists* something rather than nothing? If the universe were absolutely void, it would remain logically admissible that, supposing that something existed (which would not be the case), it would have to be equal to itself. The statement 'every x is equal to x' would be valid, but there would be no x, so the statement 'there exists an x equal to itself' would not be valid.

The passage from universal statement to assertion of existence is an exorbitant right, which the concept cannot arrogate to itself. It is not possible to elicit existence on the basis of a universal law that could be upheld just as well in absolute nothingness (consider for example the statement 'the nothing is identical to itself'). And, since no existent object can be deduced from pure thought, you cannot separate zero therein. Zermelo does not save Frege.

2.18. The existence of zero, or of the empty set, and therefore the existence of numbers, is in no way deducible from the concept, or from language. 'Zero exists' is inevitably a *first* assertion; the very one that fixes an existence from which all others will proceed. Far from it being the case that Zermelo's axiom, combined with Frege's logicism, allows us to engender zero and then the chain of numbers, it is on the contrary the absolutely inaugural existence of zero (as empty set) that ensures the possibility of separating any extension of a concept whatsoever. Number comes first here: it is that *point of* *being* upon which the exercise of the concept depends. Number, as number of nothing, or zero, sutures every text to its latent being. The void is not a production of thought, because it is from its existence that thought proceeds, in as much as 'it is the same thing to think and to be'.[16] In this sense, it is the concept that comes from number, and not the other way around.

2.19. Frege's attempt is unique in certain regards: it is not a matter of creating new intra-mathematical concepts (as will be the case in Dedekind and Cantor), but of elucidating – with the sole resource of rigorous analysis – what, among the possible objects of thought, singularises those which fall under the concept of number. In this respect, my own efforts follow along the same lines. We simply need to remove the obstacles by reframing the investigation according to new parameters. Above all, it must be shown that thought is not constituted by concepts and statements alone, but also by decisions that engage it within the *epoch* of its exercise.

3

Additional Note on a Contemporary Usage of Frege

3.1. Jacques-Alain Miller, in a 1965 lecture entitled 'Suture' and subtitled 'Elements of the logic of the signifier',[1] put forward a reprise of Frege's construction of number. His text founds a certain regime of compatibility between structuralism and the Lacanian theory of the subject. I am myself periodically brought back to this foundation,[2] albeit only on condition of disrupting it somewhat. Twenty-five years later, 'I am here; I am still here'.[3]

3.2. Miller puts the following question to Frege: '*What is it* that functions in the sequence of whole numbers?'[4] And the response to this question – a response, might I say, forcefully extorted out of Frege – is that 'in the process of the constitution of the sequence, *the function of the subject*, unrecognised,[5] is operative'.[6]

3.3. If we take this response seriously, it means that, in the last instance, in the proper mode of its miscognition, it is the function of that subject whose concept Lacan's teaching communicates to us that constitutes, if not the essence, at least the process of engenderment (the 'genesis of progression', says Miller)[7] of number.

Obviously such a radical thesis cannot be ignored. Radical, it would seem at first glance, with regard to Frege's doctrine, which dedicates a specific argument to the refutation of the idea that number might be 'subjective'[8] (although it is true that, for Frege, 'subjective' means 'caught up in representation', which obviously does not match the Lacanian function of the subject). Radical also with regard to my own thesis, since I hold that number is a form of being, and that, far from being subtended by the function of the subject, it is on the contrary on the basis of number, and especially of that first number–being that is the void (or zero), that the function of the subject receives its small share of being.

3.4. We will not undertake here to examine the importance of this text – the first great Lacanian text not to be written by Lacan himself – for the doctrine of the signifier, nor to explore what analogy it employs to illuminate the importance – at the time, still little appreciated – of all that the master taught us as to the subject's being comprised in the effects of a chain. We seek to examine exclusively what Miller's text assumes and proposes *with regard to the thinking of number as such*.

3.5. Miller's demonstration is organised as follows:

• To found zero, Frege (as we saw in **2.6**) summons to his aid the concept 'not identical to itself'. No object falls under this concept. On this point, Miller emphasises – even compounds – Frege's reference to Leibniz. To suppose that an object could be not be identical to itself, or that it could be non-substitutable for itself, would be entirely to subvert truth. In order to be true, a statement bearing upon object A must suppose the invariance of A in each occurrence of the statement, 'each time' the statement is made. The principle 'A is A' is a law of any possible truth. And reciprocally, in order that truth be saved, it is crucial that no object should fall under the concept 'not identical to itself'. Whence zero, which numbers the extension of such a concept.

• Number is thus shown to issue from the concept alone, on condition of truth. But this demonstration is consistent only *because it has been able to invoke in thought an object non-identical to itself*, even if only to discharge it in the inscription of zero. Thus, Miller writes, 'the 0 which is inscribed in the place of the number consummates the exclusion of this object'.[9]
To say that 'no object' falls under the concept 'not identical to itself' is to make this object vanish as soon as it is invoked, in this nothing the only subsisting trace of which will be, precisely, the mark zero: 'Our purpose has been,' Miller concludes, 'to recognize in the zero number the suturing stand-in for the lack'.[10]

• *What is it* that comes to lack thus? What 'object' can have as a stand-in for its own absence the first numerical mark; and support, in relation to the whole chain of numbers, the uninscribable place of that which appears only in order to vanish? What is it that insists *between numbers*? We must certainly agree that no 'object' can, even by failure or default, fall in that empty place that assigns non-self-identity. But there does exist (or here, more precisely, ek-sist) precisely that which is not object, that which is *proper* to the non-object, the object as impossibility of the object: the subject. 'The impossible object, which the discourse of logic summons as the not-identical with itself and then rejects . . . in order to constitute itself as what it is, which it summons and rejects *wanting to know nothing of it*, we name this object, in so far as it functions as the excess which operates in the sequence of numbers, the subject.'[11]

3.6. We must meticulously distinguish between that which Miller assumes from Frege and that which he deciphers in Frege's work on his own account. I will proceed in three stages.

3.7. FIRST STAGE   Miller takes as his starting point the proposition of Leibniz–Frege according to which *salva veritate*[12] demands that all objects should be identical to themselves. The whole literalisation of the real towards which Leibniz worked all his life, and to which Frege's ideography is the undoubted heir, is in fact surreptitiously assumed here. In this regard, Miller is indeed right to equate, along with Leibniz, 'identical to itself' and 'substitutable', thus denoting an equivalence between the object and the letter. For what could it mean to speak of the substitutability of an object? Only the letter is entirely substitutable for itself. 'A is A' is a principle of letters, not of objects. To be identifiable at a remove from itself, and subject to questions of substitutability, the object must fall under the authority of the letter,[13] which alone renders it over to calculation. If A is not identical at all moments to A, truth (or rather veridicality) *as calculation* collapses.

The latent hypothesis is therefore that truth *is of the order of calculation*. It is only on this supposition that, firstly, the object has to be represented as a letter; and, secondly, that the non-self-identity of the object-letter radically subverts truth. And if truth is of the order of calculation, then zero – which numbers the exclusion of the non-self-identical (the subject) – is itself nothing but a letter, the letter 0. The conclusion then follows straightforwardly that zero is the inert stand-in for lack, and that what 'drives' the sequence of numbers as a product of marks – a repetition in which is articulated the miscognition of that which insists – is the function of the subject.

More simply: if truth is saved only by upholding the principle of identity, then the object emerges in the field of truth only as a letter amenable to calculation. And, if this is the case, number can sustain itself only as the repetition of that which insists in lacking, which is necessarily the non-object (or the non-letter, which is the same thing), the place where 'nothing can be written'[14] – in short, the subject.

3.8. No one is obliged to be a Leibnizian, even if we must recognise in this philosophy the archetype of one of the three great orientations in thought, the constructivist or nominalist orientation (the other two being the transcendent and the generic).[15] As an advocate of the generic orientation, I declare that, for truth to be saved, one must precisely *abolish* those two great maxims of Leibnizian thought, the Principle of Non-Contradiction and the Principle of Indiscernibles.

3.9. A truth supposes that the situation of which it is the truth attains non-self-identity: this non-self-identity is indicated by the situation's being supplemented by an 'extra' multiple, one whose belonging or non-belonging to the situation is, however, intrinsically undecidable. I have named this supplement 'event', and it is always from an event that a truth-process originates. Now, when the undecidable event must be decided within the situation, that situation necessarily undergoes a vacillation as to its identity.

3.10. The process of a truth – puncturing the strata of knowledge harboured by the situation – inscribes itself within the situation as indiscernible infinity, which no thesaurus of established language has the power to designate.

Let's say simply that zero, or the void, has nothing in itself to do with the salvation of truth, which is at play in the 'laboured' correlation between the undecidability of the event and the indiscernibility of its result within the situation. No more so than it is possible to refer truth to the power of the letter, since the existence of a truth is precisely that to which no inscription can attest. The statement 'truth is' – far from guaranteeing that no object falls under the concept of 'not identical to itself' and that therefore zero is the number of that concept – instead allows us this threefold conclusion:

- there exists an object that has attained 'non-self-identity' (undecidability of the event);
- there exist an infinity of objects that do not fall under any concept (indiscernibility of a truth);
- number is not a category of truth.

3.11. SECOND STAGE   What is the strategy of Miller's text? And what role does number *as such* play within it? Is it really about arguing that the function of the subject is implicated – as a miscognised foundation – in the essence of number? This is undoubtedly what is stated in all clarity by the formula I have already cited above: 'In the process of the constitution of the sequence [of numbers] . . . the function of the subject . . . is operative.'[16] More precisely, only the function of the subject – that which zero, as number, marks in the place of lack, holding the place of its revocation – is capable of explaining what, in the sequence of numbers, functions as iteration or repetition: being excluded, the subject (the non-self-identical) includes itself through the very insistence of marks, incessantly repeating the 'one more step', firstly from 0 to 1 ('the 0 counts for 1', notes Miller), then indefinitely, from n to $n + 1$: 'its [the subject's – in the Lacanian sense] *exclusion* from the field of number is identified with repetition'.[17]

3.12. Other passages of Miller's text are more equivocal, indicating an analogical reading. For example: 'If the sequence of numbers, metonymy of the zero, begins with its metaphor, if the zero member of the sequence as number is only the stand-in suturing the absence (of the absolute zero) which moves beneath the chain according to the alternation of a representation and an exclusion – then what is there to stop us from recognising in the restored relation of the zero to the sequence of numbers the most elementary articulation of the subject's relation to the signifying chain?'[18] The word 'recognising' is compatible with the idea that the Fregean doctrine of number proposes a 'matrix' (the title of another article by Miller on the same question)[19] that is isomorphic with (maximum case) or similar to (minimum case), but in any case not identical to, the relation of the subject to the signifying chain. Frege's doctrine would then be a pertinent *analogon* of Lacanian logic: to which we would have no reply, since in that case Miller's text *would not be a text about number*. It would be doubly not so: firstly because it would speak, not of number, but of Frege's doctrine of number (without taking any position on the validity or consistency of that doctrine); and secondly because it would present the sequence of numbers as a didactic vector for the logic of the signifier, and not as an effective example of an implication of the function of the subject in the sequence of numbers.

3.13. This critical evasion assumes that two conditions are met: that there is, between the doctrine of number and that of the signifier, isomorphism or similarity, and not identity or exemplification; and that Miller does not account for the validity of the Fregean doctrine of number.

3.14. On this last point, where, to my mind (that is, to one who is concerned with the thinking of number as such), everything hangs in the balance, Miller maintains the suspense at every step. He speaks of 'Frege's system' without our being able to decide whether or not, in his opinion, the latter is an actually accomplished theory of number, a theory entirely defensible in essence. It is striking that at no point in this very subtle and intricate exercise are the immanent problems of 'Frege's system' ever raised – in particular, those that I highlighted above with regard to zero, the impact of Russell's paradox, Zermelo's axiom and, ultimately, the relation between language and existence. It thus remains *possible* to believe that the isomorphism signifier/number operates between, on the one hand, Lacan and, on the other, Frege reduced to a singular theory whose inconsistency is of no consequence with regard to the analogical goals pursued.

3.15. Evidently, it remains to be seen whether this inconsistency isn't, as a result, *transferred to the other pole of the analogy*, that is, to the logic of the signifier. The risk is not inconsequential, given that Miller places the latter in a founding position with regard to logic *tout court* – presumably including Frege's doctrine: 'The first [the logic of the signifier] treats of the emergence of the second [the logic of logicians], and should be conceived of as the logic of the origin of logic.'[20] But what happens if the completion of this process of origination is induced, through the theme of the subject, by a scheme (Frege's) marred by inconsistency? But this is not my problem. Given the conditions I have laid out, if the text is not about number, then we are finished here.

3.16. THIRD STAGE   There remains, however, an incontestable degree of adherence on Miller's part to a common representation of number, wherein number is conceived of as in some way intuitive, and which I cannot accept. This concerns the idea – central, since it is precisely here that the subject makes itself known as the cause of repetition – according to which number is grasped as a 'functioning', or in the 'genesis of the progression'. This is the image of a number that is 'constructed' iteratively, on the basis of that point of puncture that is denoted by zero. This dynamical theme, which would have us see number as passage, as self-production, as engenderment, is omnipresent in Miller's text. The analysis centres precisely on the 'passage' from 0 to 1, or on the 'paradox of engendering' $n + 1$ from n.

3.17. This image of number as iteration and passage precludes any orderly discussion of the essence of number. Even if we can only *traverse* the numeric domain according to certain *laws* of progression, of which succession is the most common (but not the only one, far from it), why must it follow that these laws are constitutive of the being of number? It is easy to see why *we* have to 'pass' from one number to the next, or from a sequence of numbers to its limit. But it is, to say the least, imprudent thereby to conclude that *number is defined or constituted by such passage*. It might well be (and this is my thesis) that number *does not pass*, that it is immemorially deployed in a swarming[21] coextensive with its being. And we will see that, just as these laborious passages only govern *our* passage through this deployment, in the same way it is likely that we remain ignorant of, have at the present time no use for, or no access to, the greater part of those numbers that our thought can conceive of as *existent*.

3.18. The 'constructivist' thesis, which makes of iteration, succession, passage the very essence of number, leads to the conclusion that *very few* numbers exist, since here 'exist' has no sense apart from that effectively supported by some such passage. Certainly, intuitionists assume this impoverished perspective. Even a semi-intuitionist like Borel[22] thought that the great majority of natural whole numbers 'don't exist' except as a fictional and inaccessible mass. So it might well be that the Leibnizian choice that Miller borrows from Frege is doubled by a latent intuitionist choice.

We must recognise that intuitionist logic and the logic of the signifier have more than a little in common, if only because the former expressly invokes the subject (the 'mathematician–subject') as part of its machinery. But in my opinion such a choice would represent an additional reason not to enter into a doctrine of number whose overall effect is to make the place of number, measured by the operational intuition of a subject, inexorably finite. For the domain of number is rather an ontological prescription incommensurable with any subject and immersed in the infinity of infinities.

3.19. The problem now becomes: how to think number whilst admitting, against Leibniz, that there are real indiscernibles; against the intuitionists, that number persists and does not pass; and against the foundational use of the subjective theme, that number exceeds all finitude?

4

Dedekind

4.1. Dedekind[1] introduces his concept of number within the framework of what we would today call a 'naïve' theory of sets. 'Naïve' because a theory of multiplicities is advanced that recapitulates various presuppositions about things and about thought. 'Naïve' meaning, in fact: philosophical.

Dedekind states explicitly, in the opening of his text *The Nature and Meaning of Numbers*, that he understands 'by *thing* every object of our thought';[2] and, a little later, that, when different things are 'for some reason considered from a common point of view, associated in the mind, we say that they form a system S'.[3] A system in Dedekind's sense is therefore quite simply a set in Cantor's sense. The space of Dedekind's work is not the concept (as in Frege), but, directly, the pure multiple, a collection that counts for one (as *a* system) objects of thought.

4.2. Dedekind develops a conception of number that (like Cantor's) is essentially *ordinal*. We saw (compare **2.3**) that Frege's conception was essentially *cardinal* (proceeding via biunivocal correspondences between extensions of concepts). What is the significance of this distinction? In the ordinal view, number is thought as a link in a chain, it is an element of a total order. In the cardinal view, it is rather the mark of a 'pure quantity' obtained through the abstraction of domains of objects having 'the same quantity'. The ordinal number is thought according to the schema of a *sequence*, the cardinal number, according to that of a *measurement*.

4.3. Dedekind affirms that infinite number (the totality of whole numbers, for example) *precedes*, in construction, finite number (each whole number, its successor, and so on). Thus the existence of an infinite (indeterminate) system, and then the particular existence of N (the set of natural whole numbers) form the contents of the paragraphs numbered 66 and 72 in Dedekind's text, whereas a result as apparently elementary as 'every number *n* is different from the following number *n*' comes only in paragraph 81.

Dedekind is a true modern. He knows that the infinite is *simpler* than the finite, that it is the most general attribute of being, an intuition from which Pascal had already drawn radical consequences – and was the first to do so – as to the site of the subject.

4.4. Dedekind first of all invites us to accept the philosophical concept of 'system', or any multiplicity whatsoever (compare **4.1**). The principal operator will then be, as in Frege (**2.3**), the idea of biunivocal correspondence between two systems. Dedekind, however, will make use of it in a totally different way than did Frege.

Let's note in passing that the biunivocal correspondence, bijection, is the key notion of all the thinkers of number of this epoch. It organises Frege's thought, Cantor's and Dedekind's.

4.5. Dedekind calls the function, or correspondence, a 'transformation',[4] and what we would call a bijective function or a biunivocal correspondence he calls a 'similar transformation'.[5] In any case, we are dealing with a function f which makes every element of a set (or system) S' correspond to an element (and one only) of a set S, in such a way that:

–   to two different elements s_1 and s_2 of S will correspond two different elements $f(s_1)$ and $f(s_2)$ of S';
–   every element of S' is the correspondent, through f, of an element of S.

A *distinct* (today we would say injective) function is a function that complies only with the first condition:

$$[(s_1 \neq s_2) \rightarrow (f(s_1) \neq f(s_2))]$$

Evidently, such functions can be defined 'in' a system S, rather than 'between' a system S and another system S'. Functions (or transformations) of this type make every element of S correspond to an element of S (either another element or the same one: the function could be the function of identity, at least for the element in question).

4.6. Take, then, a system S, an application f (not necessarily one of likeness or a biunivocal one) of S to itself, and s, an element of S. We will call the *chain* of the element s for the application f, the set of values of the function obtained by iterating it starting from s. So the chain of s for f is the set whose elements are: $s, f(s), f(f(s)), f(f(f(s))), \ldots$, etc.

We are not necessarily dealing here with an infinite iteration: it could very well be that, at a certain stage, the values thus obtained would repeat themselves. This is evidently the case if S is finite, since the possible values, which are the elements of S (the application f operates from S within S), will be exhausted after a finite number of stages. But it would also be the case were one to come across a value p of the function f where, for p, f is identical. Because then $f(p) = p$, and therefore $f(f(p)) = f(p) = p$. The function *halts* at p.

4.7. We will say that a system N is (this is Dedekind's expression) *simply infinite*[6] if there exists a transformation f of N within N that complies with the three following conditions:

1 The application f of N within N is a distinct application (cf. 4.5).
2 N is the chain of one of its elements, which Dedekind denotes by 1, and which he calls the *base-element* of N.
3 The base-element 1 is not the correspondent through f of any element of N. In other words, for any n which is part of N, $f(n) \neq 1$: the function f never 'returns' to 1.

We can form a simple enough image of such an N. We 'start' with the element 1. We know (condition 3) that $f(1)$ is an element of N different from 1. Next we see that $f(f(1))$ is different from 1 (which is never a value for f). But, equally, $f(f(1))$ is different from $f(1)$. In fact, the function f (condition 1) is a distinct transformation – so two different elements must correspond, through f, to different elements. From the fact that 1 is different from $f(1)$ it follows that $f(1)$ is different from $f(f(1))$. More generally, every element obtained through the iteration of function f will be different from all those that 'preceded' it. And, since N (condition 2) is nothing other than the chain thus formed, N will be composed of an 'infinity' (in the intuitive sense) of elements, all different, ordered by function f, in the sense that each element 'appears' through an additional step of the process that begins with 1 and is continued by repeatedly applying operation f.

4.8. The 'system' N thus defined is the *place of number*. Why? Because all the usual 'numerical' manipulations can be defined on the elements n of such a set N.

By virtue of the function f, we can pass without difficulty to the concept of the 'successor' of a number: if n is a number, $f(n)$ is its successor. It is here that Dedekind's 'ordinal' orientation comes into effect: function f, via the mediation of the concept of the chain, is that which defines N as the space of a total order. The first 'point' of this order is obviously 1. For philosophical reasons (compare **1.17**), Dedekind prefers a denotation beginning with 1 to one beginning with 0; '1' denotes the first link of a chain, whereas zero is 'cardinal' in its very being: it marks lack, the class of all empty extensions.

With 1 and the operation of succession it will be easy to obtain, firstly, the primitive theorems concerning the structure of the order of numbers, and then the definition of arithmetical operations, addition and multiplication. On the sole basis of the concepts of 'system' (or set) and of 'similar transformation' (or biunivocal correspondence), the 'natural' kingdom of numericality will be rediscovered.

4.9. A system N, structured by a function f which complies with the three conditions above (**4.7**) will be called 'a system of numbers', a place of the set of numbers. To cite Dedekind:[7]

> If, in the consideration of a simply infinite system N, set in order by a transformation f, we entirely neglect the special character of the elements, simply retaining their distinguishability and taking into account only the relations to one another in which they are placed by the order-setting transformation f, then are these elements called *natural numbers* or *ordinal numbers* or simply *numbers*, and the base-element 1 is called the *base-number* of the *number-series* N. With reference to this freeing the elements from every other content (abstraction), we are justified in calling numbers a free creation of the human mind.

The enthusiastic tone leaves no room for doubt: Dedekind is conscious of having, with his purely functional and ordinal engendering of 'system' S, torn number away from any form of external jurisdiction, in the direction of pure thought. This was already the tone, and these the stakes, of the 'proclamation' that appeared in the Preface to the first edition of his pamphlet: 'In speaking of arithmetic (algebra, analysis) as a part of logic, I mean to imply that I consider the number concept to be entirely independent of the notions or intuitions of space and time, that I consider it more as an immediate result of the laws of thought.' This is a text that, as will be appreciated, lends itself to a Kantian interpretation: the whole problem for modern thinkers of number is to navigate within the triangle Plato–Kant–Leibniz.[8] In defining, not 'a' number, but N, the simply infinite 'system' of numbers, Dedekind considers, with legitimate pride, that he has established himself, by means of the power of thought alone, in the intelligible place of numericality.

4.10. Informed by Frege's difficulties, which do not concern his concept of zero and of number, but the transition from concept to existence or the jurisdiction of language over being, we ask: does a system of numbers, a 'simply infinite' system N, *exist*? Or will some unsuspected 'paradoxes' come to temper, for us, Dedekind's intellectual enthusiasm?

4.11. Dedekind is evidently concerned about the existence of his system of number. In order to establish it, he proceeds in three steps:

1   Intrinsic definition, with no recourse to philosophy or to intuition, of what an infinite system (or set) is.
2   Demonstration (this, as we shall see, highly speculative) of the existence of an infinite system.
3   Demonstration of the fact that all infinite systems 'contain as a proper part a simply infinite system N'.

These three steps permit the following conclusion to be drawn: since at least one infinite system exists, and every infinite system has as a subsystem an N – a simply infinite system or 'place of number' – then this place exists. Which is to say: number exists. The idea that 'arithmetic should be a part of logic'[9] signifies that, by means of the conceptual work of pure thought alone, I can guarantee the consistency of an intelligible place of numericality, and the effective existence of such a place.

4.12. Dedekind's definition of an infinite set is remarkable. He himself was very proud of it, and with good reason. He notes that 'the definition of the infinite . . . forms the core of my whole investigation. All other attempts that have come to my knowledge to distinguish the infinite from the finite seem to me to have met with so little success that I think I may be permitted to forgo any critique of them.'[10]

This definition of the infinite systematises a remark already made by Galileo: there is a biunivocal correspondence between the whole numbers and the numbers that are their squares. Suffice to say, $f(n)$ = n^2. However, the square numbers constitute a proper part of the whole numbers (a proper part of a set is what we call a part that is different from the whole, a truly 'partial' part). It seems, therefore, in examining intuitively infinite sets, that there exist biunivocal correspondences between the sets as a whole and one of their proper parts. This part, then, has 'as many' elements as the set itself. Galileo concluded that it was absurd to try to conceive of actual infinite sets. Since an infinite set is 'as large' (contains 'as many' elements) as one of its proper parts, the statement 'the whole is greater than the part' is apparently *false* in the case of infinite totalities. Now, this statement is an axiom of Euclid's *Elements*, and Galileo did not think it could be renounced.

Dedekind audaciously transforms this paradox into the *definition* of infinite sets: 'A system S is said to be *infinite* when it is similar to a proper part of itself. In the contrary case, S is said to be a *finite* system.'[11] (Remember that, in Dedekind's terminology, 'system' means set, and the similarity of two systems means that a biunivocal correspondence exists between them).

4.13. The most striking aspect of Dedekind's definition is that it determines infinity *positively*, and subordinates the finite negatively. This is its especially modern accent, such as is almost always found in Dedekind. An infinite system has a property of an existential nature: there *exists* a biunivocal correspondence between it and one of its proper parts. The finite is that for which such a property *does not obtain*. The finite is simply that which is not infinite, and all the positive simplicity of thought hinges on the infinite. This intrepid total secularisation of the infinite is a gesture whose virtues we (inept partisans of 'finitude', wherein our religious dependence can still be read) have not yet exhausted.

4.14. The *third* point of Dedekind's approach (that every infinite system contains as one of its parts a system of type N, a place of number, see **4.11**) is a perfectly elegant proof.

Suppose that a system S is infinite. Then, given the definition of infinite systems, there exists a biunivocal correspondence f between S and one of its proper parts S'. In other words a bijective function f that makes every element of S correspond to an element of S'. Since

S′ is a *proper* part of S, there is at least one element of S that is not in the part S′ (otherwise S = S′, and S′ is not a 'proper' part). We choose such an element, and call it 1. Consider the chain of 1 for the function *f* (for 'chain' cf. **4.6**). We know that:

- *f* is a distinct (injective) transformation, or function, since it is precisely the biunivocal correspondence between S and S′, and all biunivocal correspondence is distinct;
- 1 certainly does not correspond through *f* to any other term of the chain, since we have chosen 1 from outside of S′, and *f only makes elements of S′ correspond* to elements of S. An element *s* such that *f*(*s*) = 1 therefore cannot exist in the chain. In the chain, the function never 'returns' to 1.

The chain of 1 for *f* in S is, then, a simply infinite set N: it complies with the three conditions set for such an N in **4.7** above.

We are thereby assured that, *if an infinite system S exists*, then an N, a place of number, also exists as part of that S. Dedekind's thesis is ultimately as follows: *if the infinite exists, number exists*. This point (taking account of the ordinal definition of number as the chain of 1 for a similar transformation, and of the definition of the infinite) is exactly *demonstrated*.

4.15. But does the infinite exist? There lies the whole question. This is point two of Dedekind's approach, where we see that, *for Dedekind, the infinite, upon which the existence of number depends, occupies the place which for Frege is occupied by zero.*

4.16. To construct the proof upon which henceforth all will rest (the consistency and the existence of an infinite system or set), Dedekind briskly canvasses all his initial philosophical presuppositions (the thing as object of thought). Of course, these presuppositions already quietly prop up the very idea of a 'system' (collection of anything whatsoever). But, seized by the superbly smooth surface of the subsequent definitions (chain, simply infinite set) and proofs, we had the time to let this fragility slip from our minds. We could do no better than to cite here Dedekind's 'proof' of what is put forward blithely as the 'theorem' of paragraph 66:[12]

66. *Theorem:* There exist infinite systems.

Proof: My own realm of thoughts, i.e. the totality *S* of all things, which can be objects of my thought, is infinite. For, if *s* signifies an element of *S*, then is the thought *s*′, that *s* can be an object of my thought, itself an element of S. If we regard this as transform $f(s)$ of the element s, then has the transformation f of S, thus determined, the property that the transform S' is a part of S; and S' is certainly a proper part of S, because there are elements in S (e.g. my own ego) which are different from every such thought s' and therefore are not contained in S'. Finally it is clear that, if s_1 and s_2 are different elements of S, their transforms s_1' and s_2' are also different, that therefore the transformation f is a distinct (similar) transformation. Hence S is infinite, which was to be proved.

4.17. Once our stupor dissipates (but it is of the same order as that which grips us in reading the first propositions of Spinoza's *Ethics*), we must proceed to a close examination of this proof of existence.

4.18. Some technical specifics: The force of the proof lies in the consideration of the correspondence between an 'object of my thought' and the thought 'this is an object of my thought' – that is to say, the correspondence *between a thought and the thought of that thought, or reflection* – as a function operating between elements of the set of my possible thoughts (in fact, we may as well identify a 'possible object of my thought' with *one* possible thought). This function is 'distinct' (we would now say injective), because it possesses the property (which biunivocal correspondences also possess) that two distinct elements always correspond via the function to two distinct elements. Given two thoughts whose objects distinguish them from each other, the two thoughts of these thoughts are distinct (they also have distinct objects, since they think of distinct thoughts). Consequently there is a biunivocal correspondence between thoughts in general and thoughts of the type 'thought of a thought'. Or, if you prefer, there is such a correspondence between thoughts whose object is anything whatsoever and thoughts whose object is a thought. Now this second set forms a proper part of the set of all possible thoughts, since there are thoughts which are *not* thoughts of thoughts: the striking example Dedekind gives is what he calls 'the ego'. Thus the set of all my possible thoughts, being in biunivocal correspondence with one of its proper parts, is infinite.

4.19. Dedekind's approach *is a singular combination of Descartes' Cogito and the idea of the idea in Spinoza.*

The starting point is the very space of the Cogito, as 'closed' configuration of all possible thoughts, existential point of pure thought. It is claimed (but only the Cogito assures us of this) that something like the set of all my possible thoughts *exists*.

From Spinoza's causal 'serialism' (regardless of whether or not he figured in Dedekind's historical sources) are taken both the existence of a 'parallelism' which allows us to identify simple ideas by way of their object (Spinoza says: through the body of which the idea is an idea), and the existence of a reflexive redoubling, which secures the existence of 'complex' ideas, whose object is no longer a body, but another idea. For Spinoza, as for Dedekind, this process of reflexive redoubling must *go to infinity*. An idea of an idea (or the thought of a thought of an object) *is an idea*. So there exists an idea that is the idea of the idea of the idea of a body, and so on.

All of these themes have to be in place in order for Dedekind to be able to conclude the existence of an infinite system. There must be a circumscribed 'place', representable under the sign of the One, of the set of my possible thoughts. We recognise here the soul, the 'thinking thing' as paradigmatically established by Descartes, in its existence and essence (pure thought), in the Cogito. An idea must be identifiable through its object, so that two different ideas correspond to two different objects: this alone authorises the biunivocal character of the correspondence. And, ultimately, it must be that the reflexive process goes to infinity, since, if it did not, there would exist thoughts with no correspondent through the function, thoughts for which there were no thoughts of those thoughts. This would ruin the argument, since it would no longer be established that to *every* element of the set of my possible thoughts S there corresponds an element of the set of my reflexive thoughts S'. Ultimately – above all, I would say – there must be at least one thought that is *not* reflexive, that is not a thought of a thought. This alone guarantees that S', the set of reflexive thoughts, is a *proper* part of S, the set of my possible thoughts. This time, we recognise in this fixed point of difference the Cogito as such – what Dedekind calls 'my own ego'. *That which does not allow itself to be thought as thought of a thought is the act of thinking itself, the 'I think'.* The 'I think' is non-decomposable; it is impossible to grasp it as a thought of *another* thought, since every other thought presupposes it.

It is therefore no exaggeration to say that for Dedekind, ultimately, number exists in so far as there is the Cogito as pure point of existence, underlying all reflection (specifically, there is an 'I think that I think'), but itself situated outside of all reflection. The existential foundation of the infinite, and therefore of number, is what Sartre calls the 'pre-reflexive Cogito'.

And here we discover a variant of Jacques-Alain Miller's thesis: what subtends number is the subject. The difference is that, whereas for Miller it is the 'process of engendering' of number that requires the function of the subject, for Dedekind it is the existence of the infinite as its place. The Fregean programme of the conceptual deduction of zero and the Dedekindian programme of the structural deduction of the infinite lead back to the same point: the subject, whether as insistence of lack or as pure point of existence. To the Lacanian subject can be ascribed the genesis of zero, to the Cartesian subject, the existence of the infinite. As if two of the three great modern challenges of thinking number (zero, the infinite, the downfall of the One), once the third is assumed in the guise of a theory of sets, can only be resolved through a radical employment of that great *philosophical* category of modernity: the subject.

4.20. I could simply say that, just as I am not enough of a Leibnizian to follow Frege, I am equally neither Cartesian nor Spinozist enough to follow Dedekind.

4.21. Against Dedekind's Spinozism: Far from the idea of an infinite recurrence of the thought of a thought of a thought of a thought of a thought, and so on, being able to found the existence of the place of number, *it presupposes it*. In fact, we *have* no experience of this type. Only the existence – and *consequently* the thought – of the sequence of numbers allows us to represent, and to make a numerical fiction of, a reflection which reflects itself endlessly. The very possibility of stating a 'thought' at, say, the fourth or fifth level of reflection obviously relies on an abstract knowledge of numbers as a condition. As to the idea of a reflection that 'goes to infinity', this obviously contains precisely what we are trying to demonstrate, namely the effect of infinity in thought: an effect whose only known medium is the mathematics of number.

4.22. As regards questions of *existence*, Spinoza himself made certain not to proceed as Dedekind does. He never sought to infer the existence of the infinite from the recurrence of ideas. It is, rather, precisely because he *postulated* an infinite substance that he was able to establish that the sequence that goes from the idea of a body to ideas of ideas of ideas, and so on, is infinite. For Spinoza, and he is quite justified in this, the existence of the infinite *is an axiom*. His problem is rather 'on the other side', the side of the body (or, in Dedekind's terms, that of the object). For, if there is a rigorous parallelism between the chain of ideas and the chain of bodies, then there must be, corresponding to the idea of an idea, the 'body of a body', and we are unable to grasp what the reality of such a thing might be. Dedekind evades this problem because the place of thinking he postulates assumes Cartesian closure: the corporeal exterior, the extensive attribute, does not intervene in it. But, in seeking to draw from Spinozist recurrence a conclusive (and non-axiomatic) thesis on the infinite, he produces only a vicious circle.

4.23. Against Dedekind's Cartesianism: It is essential to the proof that *every* thought can be the object of a thought. This theme is incontestably Cartesian: the 'I think' subtends the being of ideas in general as a 'material' of thought, and it is clear that there is no idea that cannot be a thinkable idea, that is to say (since we are speaking of the set of my possible thoughts) virtually actualisable as object of my thought. But obviously this excludes the possibility that 'it'[13] could think without my thinking that I think that thought, and without it being even *possible* to do so. Dedekind is Cartesian in his exclusion of the unconscious, which, since Freud, we know to think, and to think in such a way that some of its thoughts can be defined precisely as those that I cannot think. 'Unconscious thoughts' are precisely those unable, at least directly, to become objects of my thought.

More generally, it is doubtful, for a contemporary philosopher, whether *true* thoughts, those that are included in a generic procedure of truth, could ever be exposed as such in the figure of their reflection. This would be to imagine that their translation onto the figure of knowledge (which is the figure of reflection) is coextensive with them. Now the most solid idea of contemporary philosophy is precisely not to understand the process of truth except as *a gap* in knowledge. If 'thought' means: instance of the subject in a truth-procedure, then there is no thought of this thought, because it contains no knowledge. Dedekind's approach founders on the unconscious, and does not hold firmly enough to the distinction between knowledge and truth.

4.24. Descartes himself is more prudent than Dedekind. He makes certain not to infer the infinite from reflection, or from the Cogito as such. He does not consider, in proving the existence of God, the *totality* of my possible thoughts, as Dedekind does. On the contrary, he singularises *an* idea, the idea of God; his local argument might be contrasted with Dedekind's global, or set-theoretical, argument. Descartes' problem is elsewhere, it is a Fregean problem: how do we pass from concept to existence? For this, an argument positing a dispro-portion *between the idea and its place* is necessary: the idea of the infinite is without common measure with its place, which is my soul – or, in Dedekind's terms, the set of my possible thoughts; because this place, grasped in its substantial being, is finite. The singular idea of the infinite must therefore 'come from elsewhere'; it must come from a real infinity.

We can see how, in the end, Descartes' and Dedekind's positions are reversed. For Dedekind, it is the place that is infinite, because it must support reflection (the capacity of the Cogito) in its going to infinity. For Descartes, it is the exterior of the place (God) that is infinite, since the place of my thought, guaranteed in its being by the Cogito, is finite, and is therefore not *capable* of supporting alone the idea of the infinite. But, in seeking to break with the finitude of the place, Dedekind forgets that this place could well be nothing but a scene fabricated by an Other place, or that thought could well find its principle only in a *presupposition* of infinite number, of which it would be the finite and irreflexive moment.

4.25. Immanent critique: Dedekind's starting point is 'the realm of all possible objects of my thought', which he immediately decides to call system S. But *can this domain be considered as a system, that is to say, a set?* Do the 'possible objects of my thought' form *a* set, a consistent multiplicity, which can be counted as one (leaving aside the thorny question of knowing *what* carries out this accounting of my thoughts)? Isn't it rather an inconsistent multiplicity, in so far as its total recollection is, for thought itself, precisely impossible? If one admits the Lacanian identification of the impossible and the real, wouldn't the 'system' of *all* possible objects of my thoughts be the real of thought, in the guise of the impossibility of its counting-for-one? Before establishing that the 'realm of all possible objects of my thought' is an infinite system, then, we must establish that it is a system (*a* set) at all.

4.26. In the same way in which Russell's paradox comes to spoil Frege's derivation of number on the basis of the concept, the 'paradox' of the set of all sets – a descendant of the former – comes to break Dedekind's deduction of the existence of the infinite, and consequently the deduction of the existence of N, the 'simply infinite' set which is the place of number. Conceptually set out by Dedekind with impeccable inferences, the place of number does not stand the test of consistency, which is also that of existence.

4.27. Reasoning '*à la* Dedekind': Any system whatsoever (a set), grasped in abstraction from the singularity of its objects or, as Dedekind says, thought uniquely according to 'that which distinguishes' these objects (thus, their simple belonging to a system and its laws), is obviously a possible object of my thought. Consequently, within the supposed system S of all possible objects of my thought must figure, as a subsystem (subset), the system of all systems, the set of all sets. By virtue of this fact, this system of all systems is itself a possible object of my thought. Or, in simplified terms, the system of all systems is a thought.

Now, this is an impossible situation. In fact, a fundamental principle of Dedekind's demonstration has it that every thought gives rise to a thought of this thought, which is different from the original thought. So if there exists a thought of the set of all sets, there must exist a thought of this thought, *which is in S*, the set of all my possible thoughts. S is then *larger* than the set of all sets, since it contains at least one element (the thought of the set of all sets) that does not figure in the set of all sets. Which cannot be, since S *is* a set, and therefore must figure as an element *in* the set of all sets.

Or, once again: considered as a set or system, S, the domain of all the possible objects of my thought, is an element of the set of all sets. Considered in its serial or reflexive capacity, S overflows the set of all sets, since it contains the thought of that thought which is the set of all sets. S is thus at once inside (or 'smaller than') and outside (or 'larger than') one of its elements: the thought of the set of all sets. We must conclude then, excluding logical inconsistency, either that the set of all sets, the system of all systems, *is not a possible object of my thought*, even though we have just thought it; or, more reasonably, that the domain of all possible objects of my thought is not a system, or a set. But, in that case, it cannot be used to support the proof of the existence of an infinite system.

4.28. Reasoning more mathematically now: Suppose that the set of all sets exists (which implies necessarily the existence *as set* of the domain of all possible objects of my thought). Then, since it is a set, we can separate (Zermelo's axiom, **2.12**), as an existent set, all of the elements that have a certain property in common. Take the property 'not being an element of itself'. By means of separation this time, and therefore with the guarantee of existence already in place, we 'cut out' from the set of all sets, which we suppose to exist, the set of all the sets which do not belong to themselves. This set then exists, which Russell's paradox tells us is impossible (admitting the existence of the set of all sets which do not belong to themselves leads directly to a formal contradiction, cf. **2.11**). So it is impossible that the set of all sets should exist, and a fortiori that the domain of all my possible thoughts could be a set.

4.29. Dedekind's attempt ultimately fails at the same point as did Frege's: in the transition from concept to assertion of existence. And at the root of the affair is the same thing: Frege and Dedekind both seek to deduce from 'pure logic', or thought as such, not just the operational rules of number, but the fact of its existence for thought. Now, just like the empty set, or zero, *the infinite will not be deduced*: we have to *decide* its existence axiomatically, which comes down to admitting that one takes this existence, not for a construction of thought, but for a fact of Being.

The site of number, whether we approach it, like Frege, 'from below', on the side of pure lack, or, like Dedekind, 'from above', from the side of infinity, cannot be established logically, by the pressure of thought alone upon itself. There has to be a pure and simple *acknowledgement* of its existence: the Axiom of the Empty Set founds zero, and, as a result of this, the finite cardinals exist. The Axiom of Infinity founds the existence of the infinite ordinals, and from there we can return to the existence of finite ordinals. The challenges posed to the moderns by the thinking of number cannot be met through a *deduction*, but only through a *decision*. And what subtends this decision, as to its veridicality, relates neither to intuition nor to proof. It relates to the decision's conformity to that which being qua being prescribes to us. From the fact that the One is not, it follows, with regard to zero and the infinite, that nothing can be said other than: they are.

4.30. Nevertheless, we must give Dedekind immense credit for three crucial ideas.

The first is that the best approach to number is a general theory of the pure multiple, and therefore a theory of sets. This approach, an ontological one, entirely distinguishes him from the conceptual or logicist approach, as found in Frege.

The second is that, within this framework, we must proceed in 'ordinal' fashion, erecting a sort of universal series where number *will come to be grasped*. Certainly, the theory of ordinals must be removed from its overdependence on the idea of order, still very much present in Dedekind. Because, as I objected to Jacques-Alain Miller, there is no reason to presume that the being of number will be awaiting us along the ordered route that we propose to it. The concept of the ordinal must be still further ontologised, rendered less operational, less *purely* serial.

The third of Dedekind's great inspired ideas is that, to construct a modern thinking of number, a non-Greek thinking, we must *begin* with the infinite. The fact that it is vain to try to give this beginning the form of a proof of existence is ultimately a secondary matter, compared to the idea of the beginning itself. It is truly paradigmatic to have understood that, in order to think finite number, natural whole number, it is necessary *first* to think, and to bring into existence – by way of a decision that respects the historial nature of being, in so far as our epoch is that of the secularisation of the infinite (of which its numericisation is the first instance) – infinite number.

On these three points, Dedekind is truly the closest companion, and in certain respects the ancestor, of the father – still misunderstood – of the great laws of our thought: Cantor.

5

Peano

5.1. Peano's work is certainly not comparable in profundity or in novelty either to Frege's or to Dedekind's. His success lies more in the clarification of a symbolism, in the firm assurance of the connection between logic and mathematics, and in a real talent for discerning and denoting the pertinent axioms. One cannot speak of number without tackling the famous 'Peano axioms' at their source; they have become the reference text for any kind of formal introduction on the natural whole numbers.

5.2. Even though, from the very beginning of his *Principles of Arithmetic*,[1] – written, deliciously, in Latin – Peano speaks of 'questions that pertain to the foundations of mathematics', which he says have not received a 'satisfactory solution',[2] the approach he adopts is not so much a fundamental meditation as a 'technicisation' of procedures, with a view to establishing a sort of consensus on manipulation (something in which, in fact, he succeeds perfectly). This is the sense in which we ought to understand the phrase: 'The difficulty has its main source in the ambiguity of language.'[3] To expound number in the clarity of a language – an artificial clarity, certainly, but legible and indubitable – this is what is at stake in Peano's work.

5.3. In terms of its content, the approach is modelled on Dedekind's. We 'start' from an initial term, which, as with Dedekind, is not zero but one. We 'put to work' the successor function (denoted in Peano according to the additive intuition: the successor of n is written

$n + 1$). We rely heavily on induction, or reasoning by recurrence. But, whereas Dedekind, who works in a set-theoretical framework, *deduces* the validity of this procedure, in Peano it is treated purely and simply as an axiom. We decide that:

- if 1 possesses a property,
- and if it is true that, when n possesses a property, then $n + 1$ also possesses it,
- then, *all* numbers n possess the property.

Armed with this inductive principle and with purely logical axioms whose presentation he has clarified, Peano can *define* all the classical structures of the domain of whole numbers: total order and algebraic operations (addition, multiplication).

5.4. The axiom of induction, or of recurrence, marks the difference in thinking between Peano and Dedekind on the crucial issue of the infinite. Treated as a simple operational principle, recurrence actually permits legislation over an infinite totality *without making mention of its infinity*.

It is clear that there is an infinity of whole numbers. To speak of 'all' these numbers therefore means to speak of an actual infinity. But in Peano's axiomatic apparatus, this infinity is not introduced as such. The axiom of recurrence permits us, from a *verification* (1 possesses the property) and an implicative *proof* (*if* n possesses the property, *then* $n + 1$ also possesses it), to conclude that 'all numbers possess the property', without having to inquire as to the extension of this 'all'. The universal quantifier here masks the thought of an actual infinity: the infinite remains a latent form, inscribed in the quantifier without being released into thought.

Thus Peano introduces the concept of number without transgressing the old prohibition on actual infinity, a prohibition that still hangs over our thought even as the latter is summoned to its abolition by the modern injunction of being. Peano's axiomatic *evades* the infinite, or explicit mention of the infinite.

For Dedekind, on the other hand, not only the concept of the infinite, but also its existence, is absolutely crucial. Dedekind says this explicitly in a letter to Keferstein:[4]

> After the essential nature of the simply infinite system, whose abstract type is the number sequence N, had been recognized in my analysis ... the question arose: does such a system *exist* at all in the realm of our ideas? Without a logical proof of existence it would always remain

doubtful whether the notion of such a system might not perhaps contain internal contradictions. Hence the need for such proofs.

5.5. Peano does not broach questions of existence. When a system of axioms is applied to operational arrangements, we will be able, if necessary, to enquire as to that system's *coherence*; we need not speculate on the being of that which is interrogated. The vocabulary of the 'thing', or object, common to Frege and Dedekind (even if it is a matter of 'mental things' in the sense of Husserl's noematic correlate) is dropped in Peano's work, in favour of a somewhat 'postmodern' sensibility where the sign reigns. For example, he writes: 'I have denoted by signs all ideas that occur in the principles of arithmetic, so that every proposition is stated only by means of these signs.'[5] If the latent model in Dedekind and of Frege is philosophical ('philosophy as rigorous science'),[6] in Peano it is directly algebraic: 'With these notations, every proposition assumes the form and the precision that equations have in algebra . . . the procedures are similar to those used in solving equations.'[7]

The 'economy of number' proposed by Peano is an economy of signs whose paradigm is algebraic, whose transparency is consensual, and whose operational effectiveness is therefore not in doubt. He thus participates forcefully in that movement of thought, victorious today, that *wrests* mathematics from its antique philosophical pedestal and represents it to us as a grammar of signs where all that matters is the making explicit of the code. Peano prepares the way from afar – by eliminating all idea of a being of number, and, even more so, that of number *as* being – for Carnap's major theses, which reduce mathematics, treated as a 'formal language' (as opposed to empirical languages), not to a science (because according to this conception every science must have an 'object'), but to the syntax of the sciences. Peano is inscribed in the twentieth century's general movement of thought – forged, in fact, at the end of the nineteenth century – whose characteristic gesture is the destitution of Platonism, in the guise of that which had always been its bastion: mathematics, and especially the Idea of number.

5.6. We see here, as if in the pangs of its birth, the real origin of what Lyotard calls the 'linguistic turn' in Western philosophy, and what I call the reign of the great modern sophistry: if it is true that mathematics, the highest expression of pure thought, in the final analysis consists of nothing but syntactical apparatuses, grammars of signs, then a fortiori all thought falls under the constitutive rule of language.

It is certain that, for Plato, the subordination of language to 'things themselves', as dealt with for example in the *Cratylus*, has as its horizon of certitude the ontological vocation of the matheme. There is no upholding the pure empire of the sign if number, which we indicate with just a simple stroke, is, as Plato thought, a form of Being. Conversely, if number is nothing but a grammar of special signs, ruled by axioms with no foundation in thought, then it becomes probable that philosophy must be, first and foremost (as in Deleuze's reading of Nietzsche's 'diagnostics'), a thinking of the force of signs. Either truth or the arbitrariness of the sign and the diversity of syntactical games: this is the central choice for contemporary philosophy. Number occupies a strategic position in this conflict, because it is simultaneously the most generalised basis of thought and that which demands most abruptly the question of its being.

Peano's axiomatic, poor in thought but strong in its effects, a grammar which subdues number, the organising principle of an operational consensus, a deft mediation of the infinite into the finitude of signs, represents something of a lucky find, a gift, for modern sophistry.

5.7. Every purely axiomatic procedure introduces *undefined* signs, which can only be presented by codifying their *usage* in axioms. Peano is hardly economical with these 'primitive' signs: there are four, in fact (you are reminded that set theory has recourse to one single primitive sign $\in$, belonging, which denotes presentation as such):

> Among the signs of arithmetic, those that can be expressed by other signs of arithmetic together with the signs of logic represent the ideas that we can define. Thus, I have defined all signs except four ... If, as I think, these cannot be reduced any further, it is not possible to define the ideas expressed by them through ideas assumed to be known previously.[8]

These four irreducible signs are:[9]

1 The sign N, which 'means *number (positive whole number)*'.
2 The sign 1, which 'means *unity*'.
3 The sign $a + 1$, which 'means *the successor of* a'.
4 The sign =, which 'means *is equal to*'.

Peano thus explicitly renounces all definition of number, of succession, and of 1. (The case of the sign = might be treated separately: it is in point of fact a logical sign, not an arithmetical one. Peano himself writes: 'We consider this sign as new, although it has the form of a sign of logic.')[10] Evidently this is the price to be paid for operational transparency. Where Frege musters all thought towards attempting to *understand* the revolutionary statement 'zero is a number', Peano simply notes (it is the first axiom of his system): 1 ∈ N, a formal correlation between two undefined signs that 'means' (but according to what doctrine of meaning?) that 1 is a number. Where Dedekind generates the place of number as the space of possible employment, or the really existing infinite chain, of a biunivocal function, Peano notes:[11] $a \in N \rightarrow a + 1 \in N$, an implication that involves three undefined signs, and which 'means' that, if a is a number, its successor is also a number. The force of the letter is here at the mercy of meaning. And the effect is not one of obscurity, but rather one of an excessive limpidity, a cumbersome levity of the trace.

5.8. In the poem, the obscure is born of that which, as a breaking open of the signifier at the limits of language, disseminates the letter. In Peano's pure axiomatic, the retreat of sense issues from the fact that the force of the letter is turned back upon itself, and that it is only *from outside* that thought can come to it. Peano wishes to put off any confrontation with the latent poem the absence of which number – astral figure of being ('cold with neglect and disuse, a Constellation')[12] – unfailingly instigates and the effect of which Frege and Dedekind unconsciously preserve in the desperate attempt to conjure forth into Presence now zero, now the infinite.

5.9. Peano's axiomatic is a shining success story of the tendency of our times to see nothing in number except for a network of operations, a manipulable logic of the sign. Number, Peano thinks, *makes signs* about the sign, or is the Sign of signs.

From this point of view, Peano is as one with the idea that the universe of science reaches its apex in the forgetting of being, homogenous with the reabsorption of numericality into the unthought of technical will. Number is truly *machinic*. Thus it can be maintained that the success of Peano's axiomatic participates in the great movement that has *given up* the matheme to modern sophistry, by unbinding it from all ontology and by situating it within the sole resources of language.

5.10. It will be a great revenge upon this operation to discover, with Skolem and then Robinson,[13] the semantic *limits* of the grammar of signs to which Peano had reduced the concept of number. We know today that such an axiomatic admits of 'non-standard' models, whose proper being differs greatly from all that we intuitively understand by the idea of natural whole number. So that Peano's system admits of models where there exist 'infinitely large' numbers, or models whose type of infinity exceeds the denumerable. Peano arithmetic is susceptible to 'pathological' interpretations; it does not have the power to establish a univocal thought within the machinism of signs. Every attempt to reduce the matheme to the sole spatialised evidence of a syntax of signs runs aground on the obscure prodigality of being in the forms of the multiple.

5.11. The essence of number will not be spoken, either as simple force of counting and of its rules, or as sovereignty of graphisms. We must pass into it through a meditation on its being.

N is not an 'undefined' predicate, but the infinite place of exercise of that which succeeds the void (or zero), the existential seal which strikes *there* where it[14] insists on succeeding.

What 'begins' is not the 1 as opaque sign of 'unity', but zero as suture of all language to the being of the situation whose language it is.

Succession is not the additive coding of a + 1, but a singular disposition of *certain* numbers which *are* successors rather than their succeed*ing*, and which are marked in their being by this disposition. We must know also that zero and the infinite are precisely *that which does not succeed*, and that they are so in their very being, in different ways; although both are located, by virtue of this fact, on the shores of a Nothingness.

Number is neither that which counts, nor that with which we count. This regime of numericality organises the forgetting of number. To *think* number requires an overturning: it is because it is an unfathomable form of being that number prescribes to us that feeble form of its approximation that is counting. Peano presents the inscription of number, which is *our* infirmity, our finitude, as the condition of its being. But there are more things, infinitely more, in the kingdom of Number, than are dreamt of in Peano's arithmetic.

6

Cantor: 'Well-Orderedness' and the Ordinals

6.1. The ordinals represent the general ontological horizon of numericality. Following the elucidation of the concept of the ordinal, with which we shall presently occupy ourselves, this principle will govern everything that follows, and it is well said that in this sense Cantor is the true founder of the contemporary thinking of number. In fact, Cantor[1] considered that the theory of ordinals constituted the very heart of his discovery. Today, the *working mathematician*, for whom it suffices *that there are* sets and numbers and who does not worry at all about what they *are*, thinks of the ordinals rather as something of a curiosity. We must see in this mild disdain one of the forms of submission of the mathematician, in so far as he or she is exclusively *working*, to the imperatives of social numericality. Specialists in mathematical logic or set theory are doubtless an exception, even if they themselves often regret this exception: in spite of themselves, they are closest to the injunction of Being, and for them the ordinals are essential.

6.2. I have said, in connection with Dedekind, that, in our present philosophical discourse, we must assume as complete an 'ontologisation' of the ordinals as possible. The presentation of this concept by Dedekind or Cantor relates it essentially to the notion of well-orderedness – something still very close to a simple serial or operational intuition of number.

6.3. Every schoolboy knows that, given two different whole numbers, one of them is larger and the other smaller. And he knows also that, given a 'bunch' of numbers, there is one and one only that is the smallest of the bunch.

From this serial knowledge, if one abstracts out its general properties, the concept of the well-ordered set can be developed.

6.4. A 'well-ordered' set is a set for which:

- between the elements of the set, there is a relation of *total order*; given two elements, e and e', if < denotes the order-relation, then either $e < e'$, $e' < e$, or $e = e'$; no two elements are 'non-comparable' by this relation;
- given any non-empty part of the set so ordered, there is a *smallest element* of this part (an element of this part that is smaller than all the others). If P is the part considered, there exists p, which belongs to P and for which, for every other p' belonging to P, $p < p'$ This element p will be called the minimal element of P.

If an element p is minimal for a part P, it *alone* possesses that property. For, if there were another, a p' different from p, then, because the order is total, either $p < p'$ and p' would not be minimal, or $p' < p$ and p would not be minimal. So we can speak without hesitation of the 'minimal element' of a part P of a well-ordered set.

We can see that the general concept of the well-ordered set is merely a sort of extrapolation from what the schoolboy observes in the most familiar numbers: the natural whole numbers.

6.5. A good image of a well-ordered set is as follows. Let E be such a set. 'Start' with the smallest element of E, which, given condition 2 above, must exist. Call this element 1. Consider the part of E obtained by removing 1, the part $(E - 1)$. It too has a minimal element, which comes in a certain sense *straight after 1*. Call this element 2. Consider the part of E obtained by removing 1 and 2 to be the part $(E - (1,2))$. It has a minimal element, call it 3, and so on. A well-ordered set presents itself like a chain, so that every link of the chain follows ('follows' meaning: comes just after in the relation of total order) only one other, well determined (it is the minimal element *of what remains*).

6.6. Cantor's stroke of genius was to refuse to limit this image to the finite, and thereby to introduce infinite numerations. He had the following idea: If I suppose the existence – *beyond* that sequence $1,2,3, \ldots, n, n + 1, \ldots$ – of a whole number which is the 'first' well-ordered set, the matrix of all others, an 'infinite ordinal number' ω, and declare it larger than all the numbers that precede it, then what prevents me from *continuing*? I can very well treat ω as the minimal element of a well-ordered set that comes in some sense *after* the set of all the whole numbers. And I can then consider the 'numbers' ω + 1, ω + 2, . . . , ω + n, . . . , etc. I will arrive eventually at ω + ω, and will continue once again. No stopping point is prescribed to me, so that I have a sort of total series, each term of which is the possible measure of every existent sequence. This term indicates to me that, *however many came before it*, it *numbers* every series of the same length.

6.7. Allow me to call *ordinal* the measure of the length of a well-ordered set, from its minimal element to its 'end'. The 'entire' sequence of ordinals would then provide us with a scale of measurement for such lengths. Each ordinal would represent a possible structure of well-orderedness, determined by the way in which the elements succeed each other, and by the total number of these elements. This is why we say that an ordinal, whether finite (the ordinals which come before ω, and which are quite simply the natural whole numbers) or infinite (those ordinals which come after ω), numbers a 'type of well-orderedness'.

6.8. To give a technical grounding for this idea, we will consider the class of well-ordered sets that are isomorphic to one of the sets among them (and therefore isomorphic to each other). What should we understand by this?

Take two well-ordered sets, E and E', < the order-relation of E, and <' the order-relation of E'. I will say that E and E' are isomorphic if there exists a biunivocal correspondence f (cf. **4.5**) between E and E', such that, when $e_1 < e_2$, in E, then $f(e_1) <' f(e_2)$ in E'.

We can see that f projects the order of E into the order of E', and, what's more, since f is biunivocal, there are 'as many' elements in E' as in E. We can therefore say that E and E', considered strictly from the point of view of their well-orderedness, and abstracted from the singularity of their elements, are identical: the 'morphism' (form) of their well-orderedness is 'iso' (the same), as the correspondence f assures us.

In fact, each class of well-ordered sets isomorphic to each other represents *a* well-orderedness, that well-orderedness common to all sets of that class. It is *this* well-orderedness that can be represented by an ordinal.

Thus an ordinal is the mark of a possible figure (a form, a morphism) of well-orderedness, isomorphic to all the sets that take that form. An ordinal is *the number or the cipher of a well-orderedness*.

6.9. This conception, already moving strongly in the direction of determining a *horizon* of being for all number in the form of a universal scale of measurement for forms of well-orderedness, nevertheless presents some serious difficulties; the first among them technical, the remainder philosophical.

6.10. The technical difficulties are three in number, three questions which must be answered:

1   Which is the *first* term in the total series of ordinals, the initial link that 'anchors' the whole chain? This is the conceptual question of zero or the empty set, which alone is able to number sequences of no length, sequences with no elements, the well-orderedness that orders *nothing*. This is the question that caught out Frege.
2   What exactly is the procedure of thought that allows us to suppose a beyond of the sequence of finite whole numbers? What is the gesture by which we *pass beyond* the finite, and declare ω, the first ordinal which will not be a natural whole number, the first mark of a well-orderedness that describes the structure of a non-finite set? This is the existential question of the infinite, upon which Dedekind foundered.
3   Does the universal series of ordinals – the scale of measurement of all length, whether finite or infinite, the totality of specifications of well-orderedness – *exist* in the set-theoretical framework? Isn't it – like the 'system of all the possible objects of my thought' introduced by Dedekind – an *inconsistent* totality, one that thought *cannot* take as one of its possible objects? This is the question of counting for one an 'absolute' totality. It is thus the problem of the defection of the One as soon as we claim to 'count' the universe of discourse.

And so, once again, we find ourselves returned to the three challenges of the modern thinking of number: zero, the infinite and the non-being of the One.

6.11. It rapidly turns out that the third problem admits of no positive solution. Something that was at one time put forward as a 'paradox', the Burali-Forti paradox, can actually be *proved*: the ordinals *do not form a set*, they cannot be collected in a multiple that can be counted for one. The idea of 'all' the ordinals is inconsistent, impossible; it is, to this extent, the real of the horizon of the being of number.

This proof is very closely related to that which refutes Dedekind's attempt to prove the existence of an infinite set (compare **4.28**): the set of 'all' the ordinals must itself be an ordinal, and thus it would be *inside* itself (since it is a set of *all* the ordinals) and *outside* itself (since it is not counted in the sequence it totalises). We are therefore prohibited to speak of a 'set of ordinals' with no further qualification. Which is precisely to say: 'being an ordinal' is a property *with no extension*. It is possible to *confirm* that a certain object is an ordinal (possesses the property), but not to *count for one* all the objects that have this property.

6.12. I have said enough, in my critique of Frege and Dedekind, for the treatment of problems 1 and 2 (**6.10**) to be anticipated: the existence of zero, or the empty set, and the existence of an infinite set can in no way be deduced from 'purely logical' presuppositions. They are axiomatic decisions, taken under the constraints of the historial injunction of being. The world of modern thought is nothing other than the effect of this injunction. Beginning in the Renaissance, by way of a rupture with the Greek cosmos,[2] it became *necessary*, in order to be able to think at all in accordance with our pre-understanding of ontological exigency, to assume:

- that the proper mode under which every situation 'that is' is sutured to its being is not Presence, the dehiscence of that which pro-poses itself within its limits, but pure subtraction, the unqualifiable void. In that form of being which is number, this can be stated as follows: 'zero exists', or, in a style more homogenous with Cantor's ontological creation: 'a set exists which has no elements';
- that, in their quasi-totality, and by way of a rupture with the mediaeval tradition which reserves this attribute for God alone, situation-beings are infinite; so that, far from being a predicate whose force is that of the sacred, the infinite is a *banal* determination of being, such as it proffers itself as pure multiplicity under the law of a count-for-one. In that form of being which is number, this can be stated as follows: 'an infinite set exists'; or, more technically: 'an ordinal exists which is not a natural whole number'. Or, in other words, 'ω exists'.

6.13. One had to wait practically until the beginning of the twentieth century before these decisions relating to zero and to the infinite would be recognised in themselves (under the names of the Axiom of the Empty Set and the Axiom of Infinity), although they had been operative in thought for three hundred years. But this is not surprising. We can observe a veritable *philosophical* desperation constantly putting these imperatives into reverse, whether through the intellectual dereliction of the theme of finitude or through nostalgia for the Greek ground of Presence. It is true that, when we are dealing with pure declarations, decided in themselves, these declarations exhibit the fragility of their historicity. No *argument* can support them. What's more, certain truth procedures, in particular politics, art and love, are not yet *capable of sustaining* such axioms, and so in many ways *remain Greek*. They cling to Presence (art and love), continually recusing the statement 'zero is the proper numeric name of being' in order to give tribute to the obsolete rights of the One. Or (politics) they manage finitude, corroding day after day the statement 'the situation is infinite', in order to valorise the corrupted authority of practicalities.

6.14. The two axioms of the void and of the infinite structure the entire thinking of number. The pure void is that which supports *there being* number, and the infinite, that by which it is affirmed that number is the measure of the thinking of *every* situation. The fact that this is a matter of axioms and not of theorems means that the existence of zero and of the infinite is prescribed to thought by being, in order that thought might exist *in the ontological epoch* of such an existence.

In this sense, the current strength of reactive, archaic and religious wills is necessarily accompanied by an irremediable opacity of number – which, not ceasing to rule over us, since this is the epochal law of being, nevertheless becomes unthinkable for us. Number may exist as form of being but, as a result of the total secularisation of the void and of the infinite, thought can no longer exist in the form and with the force that the epoch prescribes to it. So number will now manifest itself, without limit, as tyranny.

6.15. The principal philosophical difficulty of the Cantorian concept of the ordinals is as follows. In the presentations which bind it to the concept of well-orderedness, the theory of ordinals rather seems to 'generalise' the intuition of natural whole number that allows us to *think* the being of number. It draws its authority from that which it claims to elucidate. The idea of well-orderedness in effect does not so much found the concept of number as deduce it from the lacunary and finite experience of numerical immediacy, which I incarnated (in **6.3**) in the sympathetic figure of the schoolboy.

If we truly wish to establish the being of number as the form of the pure multiple, to remove it from the schoolroom (which means also to subtract the concept from its ambient numericality), we must distance ourselves from operational and serial manipulations. These manipulations, so tangible in Peano, project onto the screen of modern infinity the quasi-sensible image of our domestic numbers, the 1, followed by 2, which precedes 3, and then the rest. The establishing of the correct distance between thought and countable manipulations is precisely what I call the ontologisation of the concept of number. From the point at which we presently find ourselves, it takes on the form of a most precise task: the ontologisation of the 'universal' series of the ordinals. To proceed, we must abandon the idea of well-orderedness and think ordination, ordinality, in an intrinsic fashion.

It is not as a measure of order, nor of disorder, that the concept of number presents itself to thought. We demand an immanent determination of its being. And so for us the question now formulates itself as follows: which predicate of the pure multiple, that can be grasped outside of all serial engenderment, founds numericality? We do not want to count; we want to think the count.

2

Concepts: Natural Multiplicities

7

Transitive Multiplicities

7.1. What permits the abandonment of every *primitive* bond between number and order or seriality is the concept of the transitive set. Only this structural – and essentially ontological – operator enables an intrinsic determination of number as a figure of natural being. In virtue of it, we are no longer trapped in the quandaries of the deduction of the concept (Frege), of the subject as causality of lack in serial engenderment (Miller), of the existence of the infinite (Dedekind), or of the 'schoolboy' intuition of well-orderedness (Cantor).

7.2. Although this concept might seem at first glance rather mysterious, its lack of relation to any intuitive idea of number is to my eyes a great virtue. It proves that in it we grasp something that breaks the circle of an ontological elucidation of number entirely transparent in its pure and simple presupposition. We have seen that this circle recurs in Frege and in Dedekind, and that the Cantorian conception of ordinals as types of well-orderedness is still compliant with it. But we shall see that the legitimacy *for philosophical thought* of the concept of transitivity leaves no room for doubt.

7.3. To understand what a transitive set is, it is essential to penetrate the distinction – of which it would not be an exaggeration to say that it supports all post-Cantorian mathematics – between an element's *belonging* to a set and the *inclusion* of a part. This distinction is rudimentary, but it implies such profound consequences that for a long time it remained obscure.

7.4. A set is 'made out of elements', is the 'collection' (in my language, the count-for-one) of its elements.

Take the set E, and let e be one of the elements from which it 'makes' a set: we denote this by $e \in E$, and we say that e belongs to E, $\in$ being the sign for belonging.

If you now 'gather together'[1] *many* elements of E, they form a part of E. Taking E′ as the set of these elements, E′ is a part of E. This is denoted by $E' \subset E$, and we say that E′ is included in E, $\subset$ being the sign of inclusion.

Every element of a part E′ of E is an element of E. In fact this is the definition of a part: E′ is *included* in E when all the elements that *belong* to E′ *also belong* to E. So we see that inclusion is defined in terms of belonging, which is the only 'primitive' sign of set theory.

The classic (misleading) image is drawn like this:

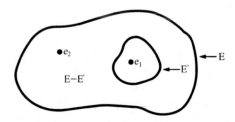

In it we can see that E′ is a part of E, that e_1 is at once (as is every element of E′) an element of E′ and an element of E, and that e_2 is an element of 'the whole' E, but not of the part E′. We also say that e_2 belongs to the *difference* of E and E′, which is denoted by E−E′.

7.5. Is it possible for an element that *belongs* to set E also to be a part of that set, also to be *included*? This seems totally bizarre, above all if we refer to the image above. But this sentiment misses the most important point, which is that an element of a set can obviously be itself a set (and even that this is always the case). Consequently, if e belongs to E, and e is a set, the question arises whether an element of e is or is not, in its turn, an element of E. If all the elements of e are also elements of E, then e, which is an element of E, is also a part of E. It belongs to E *and* is included in E.

7.6. Suppose for example that V is the set of living beings. My cat belongs to this set. But a cat is composed of cells, which one might say are themselves all living beings. So my cat is at once *a* living being and *a set* of living beings. He belongs to V (qua one, *this* living cat), and he is a part of V – he is included in V (qua group of living cells).

7.7. Forget cats. Consider the three following 'objects':

- the object e_1;
- the object e_2;
- the object which is the 'gathering together' of the first two, and which we denote by (e_1,e_2). This is called the *pair* of e_1 and e_2.

Form a set from these three objects. In the same way, we denote it by: $(e_1,e_2,(e_1,e_2))$. This is called the *triplet* of e_1 and e_2 and the pair (e_1,e_2). We will denote it by T. Note that the three elements that *belong* to this triplet are e_1, e_2, and (e_1,e_2).

Since e_1 and e_2 belong to T, if I 'gather them together', I obtain a *part* of T. Thus, the pair (e_1,e_2), which is the 'gathering together' of these two elements of T, is *included* in T. But in addition we can see that it is an element of it, that it also *belongs* to it. Thus we have constructed a very simple case of a set of which an element is also a part. In set T, the pair (e_1,e_2) is simultaneously in a position of belonging and of inclusion.

7.8. We know, from a famous theorem of Cantor's, that there are *more* parts than elements in any set E whatsoever. This is what I call the excess of inclusion over belonging, a law of being qua being whose consequences for thought are immense, since it affects the fundamental categories that inform the couplets One/Multiple and Whole/Part. It is therefore impossible that every part should be an element, that everything that is included should also belong: there are *always* parts that are not elements.

But we can put the question from the other direction: since we can see that it is possible in certain cases (for example my cat for the set V of living beings, or the pair (e_1,e_2) for our triplet T) for an element to be a part, is it possible for *all* elements to be parts, for everything that belongs to the set to be included? This is not the case for T: the element e_1 taken alone, for example, is not a part of T.

Can we produce a non-empirical example (because my V, my cat and its cells are rationally suspect) of a set all of whose elements would be parts?

7.9. Let's retrace our steps a little, back to the empty set. We have proposed (in **2.18**) the axiom 'a set exists which has no elements', that is, a set to which nothing belongs. We are going to give to this set, the 'empty' rock of the whole edifice of multiple–being, a proper name, the name '0'.

The following, extremely subtle, remark must be made: *the empty set is a part of every set*; 0 is included in E whatever E might be. Why? Because, if a set F *is not* a part of E, it is because there are elements of F that are not elements of E (if every element of F is an element of E, then by definition F is a part of E). Now 0 has no elements. So, it is impossible for it *not to be* a part of E. The empty set is 'universally' included, because nothing in it can prevent or deny such inclusion.

To put it another way: to demonstrate that F is not a part of E requires that we pick out, *within* F, at least one element: that element which, not being an element of E, proves that F cannot be included 'entirely' within E. Now the void does not tolerate any differentiation of this sort. It is in-different, and, because of this, it is included in every multiplicity.

7.10. Consider the two following 'objects':

- the empty set, 0;
- the set whose one and only element is the empty set, which is called the *singleton* of the empty set, and is denoted by (0).

Note well that this second object is *different* from the empty set itself. In fact, the empty set has *no* elements, whereas the singleton has one element – precisely, the empty set. The singleton of the void 'counts for one' the void, whereas the empty set does not count anything (this indicates a subtle distinction between 'does not count anything', which is what 0 does, and 'counts nothing', which is what (0) does. Plato already played on this distinction in the *Parmenides*).

7.11. An additional remark as regards singletons (singletons 'in general', not the particular singleton of the empty set): Take a set E and one of its elements *e* (so *e* ∈ E). The singleton of *e*, written (*e*), is a *part* of E: (*e*) ⊂ E.

What is the singleton of *e*, in fact? It is the set whose unique element is *e*. Consequently, if *e* is an element of E, 'all' the elements of the singleton (*e*) – namely the single element *e* – are elements of E, and so (*e*) is included in E.

7.12. 'Gather together' our two objects, the empty set denoted by 0 and the singleton of the empty set, denoted by (0). We obtain the pair (0, (0)), which we will denote by D. This time, the *two* elements of the pair D are also parts; everything that belongs to D is also included in D. In fact, the first element, 0, the empty set, is included in any set whatsoever (see **7.9**). Specifically, it is a part of the pair D. But, what's more, since 0 is an *element* of D, its singleton (0), is a *part* of D (**7.11**). But (0) is precisely the second element of D. Thus this element is also included in D. The set D is such that every element of it is also a part; everything that belongs to D is included in D.

7.13. As predicted by Cantor's theorem, there are parts of D that are not elements of D. For example, the singleton of the element (0) of D is a part of D, as is every singleton of an element (**7.11**). We can write this 'singleton of the singleton' as ((0)). Now, this object is *not* one of the two elements of D.

7.14. An important definition: we say that a set T is *transitive* if it is like the set D that we have just built: if all of its elements are also parts, if everything that belongs to it is also included in it, if, wherever it is the case that $t \in T$, it is also the case that $t \subset T$.

7.15. Transitive sets *exist*, without a doubt. Perhaps V, the set of living beings; certainly the set (0,(0)), which is transparent, translucent even, constructed as it is from the void (the pair of the void and the singleton of the void, the void as such and the void as one).

7.16. Modernity is defined by the fact that the One is not (Nietzsche said that 'God is dead', but for him the One of Life took the place of the deceased). So, for we moderns (or 'free spirits'), the Multiple-without-One is the last word on being qua being. Now the thought of the pure multiple, of the multiple considered in itself, without consideration of what it is the multiple of (so: without consideration of any *object* whatsoever), is called: 'mathematical set theory'. Therefore every major concept of this theory can be understood as a concept of modern ontology.

What does ontology discover in the concept of the transitive set?

7.17. Belonging is an ontological function of *presentation*, indicating *that which* is presented in the count-for-one of a multiple. Inclusion is the ontological function of *representation*, indicating multiples re-counted as parts in the framework of a representation. A most important problem (the problem of the *state of a situation*) is determined by the relation between presentation and representation.

Now, a transitive set represents the *maximum possible* equilibrium between belonging and inclusion, the element and the part, $\in$ and $\subset$. Transitivity thus expresses the superior type of ontological stability; the strongest correlation between presentation and representation.

There is always an excess of parts over elements (Cantor's theorem), there always exist parts of a set which are not elements of that set. Thus we obtain the maximal correspondence between belonging and inclusion precisely when *every* element is a part: when the set considered is transitive.

This strong internal frame of the transitive set (the fact that everything that it presents in the multiple that it is, it represents a second time in the form of inclusion), this equilibrium, this maximal stability, has led me to say that transitive sets are 'normal', taking 'normal' in the double sense of non-pathological, stable, strongly equilibriated, that is to say: not exposed to the disequilibrium between presentation and representation, a disequilibrium whose effective form is the evental caesura; and submitted to a norm, that of a maximally extended correspondence between the two major categories of ontological immanence: belonging and inclusion.

7.18. The concept of transitive multiplicity will constitute the *normal* basis for the thinking of number. Transitivity is at once that which makes of number a section[2] taken from the equilibrated fabric of being and that which provides the norm for this section.

8

Von Neumann Ordinals

8.1. Let's consider more closely set D, introduced in **7.12**, written as (0,(0)), which is the pair of the void and the singleton of the void.

We know that set D is transitive: its two elements, 0 and (0), are also parts of D. We can make a further remark here: *these two elements are also transitive sets.*

- That (0) is transitive is self-evident: the only element of the singleton (0) is 0. Now, 0 is a 'universal' part included in every set, and, in particular, it is included in the set (0). So the unique element of (0) is also a part of (0), and consequently (0) is a transitive set.
- That 0, the empty set, is transitive results from its negative 'porosity' to every property, which already makes it a part of any set whatsoever (compare **7.9**): a transitive set is one all of whose elements are also parts. Thus a set that is not transitive has at least one element that is not a part. Now 0 has no elements. So it cannot *not* be transitive. And, so, it is.

With our set D we have constructed not only a transitive set, but a transitive set of transitive sets: this transitive set 'gathers together' transitive sets. Both 0, (0) and their pair (0,(0)), are transitive.

8.2. A truly fundamental definition: *A set is an ordinal (in von Neumann's sense)*[1] if it is like D, that is, if it is transitive and all of its elements are transitive.

8.3. This definition completes the technical part of the ontologisation of the concept of the ordinal. We are no longer dealing with well-orderedness, with the image of the sequence of natural whole numbers, or with an operational status. Our concept is purely immanent. It describes a certain internal structural form of the ordinal, a form that connects together in a singular fashion the two crucial ontological operators belonging and inclusion, $\in$ and $\subset$.

Set D, which we have used as an exemplary case, is therefore an ordinal. We can lift a corner of the veil on its identity: it is *the number Two*. Moreover, this Two allows us to affirm that von Neumann ordinals *exist*.

8.4. Before deploying this new concept of the ordinal, let's begin with a first examination of the status of its definition and of the reasons why the ordinals constitute the absolute ontological horizon of *all* numbers.

8.5. I have indicated (**7.16**) that a transitive set is the ontological schema of the 'normal' multiple. Taking into account the fact that the excess of representation over presentation is irremediable, transitivity represents the maximal equilibrium between the two.

Now, not only is an ordinal transitive, but all of its elements are also transitive. An ordinal disseminates to the interior of a multiple that normality which characterises it. It is a normality of normalities, an equilibrium of equilibria.

A truly remarkable property results from this, which is that *every element of an ordinal is an ordinal*.

Take an ordinal[2] W, and an element of that ordinal x (so that $x \in W$). W being an ordinal, all of its elements are transitive, so x is transitive. For the same reason (the ordinality of W) W is itself transitive, so x, an element of W, is also a part of W: $x \subset W$. As a result, all the elements of x are elements of W. And, just as all the elements of W are transitive, the same follows for all the elements of x. The set x is thus a transitive set all of whose elements are transitive: it is an ordinal.

8.6. If transitivity is a property of stability, this time we discover a complementary property of *homogeneity*: that which makes up the internal multiple of an ordinal, the elements belonging to it, are all ordinals. An ordinal is the count-for-one of a multiplicity of ordinals.

Because of this homogenous and stable 'fabric' of ordinal multiplicity, I have been led to say that ordinals are the ontological schema of the *natural* multiple. I call 'natural' (by way of opposition to multiplicities that are unstable, heterogeneous, *historical*, and which are thus exposed to the evental caesura) precisely that which is exemplified by the underlying multiple–being as thought by mathematics: a maximal consistency, an immanent stability without lacuna, and a perfect homogeneity, in so far as that of which this multiple–being is composed is *of the same type* as itself.

We therefore posit, once and for all, that an ordinal is the index of the being of a natural multiplicity.

8.7. If it is true that the ordinals constitute the great ontological 'ground' of number, then we can also say that number is a figure of natural being, or that number *proceeds from Nature*. With the caveat, however, that 'Nature' refers here to nothing sensible, to no experience: 'Nature' is an ontological category, a category of the thought of the pure multiple, or set theory.

8.8. Must we say simultaneously that ordinals 'are numbers'? Such would indeed be the idea of Cantor, who thought to achieve by way of the ordinals an infinite prolongation of the sequence of whole numbers. But for us, who have yet to propose any concept of number, this would be begging the question. We will see, after having defined what I call Number (the capitalisation is not for the sake of majesty, but to designate a concept that subsumes all species of number, known or unknown), that the ordinals, though playing a decisive role in this definition, are only the *representable* amongst numbers, in the numerical swarming which being lavishes *on the ground of Nature*. The ordinals will thus be at once the instrument of *our* access to number, of our thinking of number, *and*, albeit lost in a profusion of Numbers that exceeds them in every way, they will be representable or figurable as themselves, too, being Numbers.

8.9. The empty set, 0, is an ordinal. We have seen above that it is transitive (**8.1**). It follows that all its elements are also transitive: having *no* elements, how could it have an element that *was not* transitive? Contrary to all intuition, zero, or the void, is a *natural* ontological donation. The void, which sutures all language and all thought to being, is also the point of nature where number is anchored.

8.10. Von Neumann ordinals have two crucial properties:

1   They are totally ordered by the fundamental ontological relation belonging, the sign of multiple–presentation. That is to say that,

given two ordinals W_1 and W_2, either the former belongs to the latter ($W_1 \in W_2$), or the other way around ($W_2 \in W_1$), or they are identical ($W_1 = W_2$).

2 They obey a principle of minimality: given any property P whatsoever, *if* an ordinal possesses this property, *then* there exists a smallest ordinal to possess it. Order is always belonging: if you have an ordinal W such that it possesses the property P (if the statement P(W) is true), then there exists an ordinal W_1 which has the property and which is the smallest to have it (if $W_2 \in W_1$, W_2 *doesn't have* the property).

These two properties are natural. The first expresses the universal intrication of those stable and homogenous multiplicities that are natural multiplicities (see **8.6**): thought in their being, two natural multiples – two ordinals, then – cannot be independent. Either one is in the presentation of the other, or vice versa. Nature does not tolerate indifference or disconnection. The second property expresses the 'atomic' or, if you like, 'quantum' character of nature. If a property applies to some natural multiple, then there is always a natural multiple that is the minimal support of that property.

Taken together, these two properties reunite the global status of nature with its local status. Even though *Nature*[3] does not exist (there is no set of all the ordinals, see **6.11**), there is a sort of unity of plan, of global interdependence, between natural multiples: the presentation of which they are the schema is always 'embedded'. And, although there are no unique and indiscernible components of nature like the Ancients' atoms (unless one considers the void as such), there is an exceptional local point for every property that obtains for the 'regions' of nature: the minimal support of this property.

This articulation of the global and the local lends its ontological framework to every Physics.

8.11. The two crucial properties (total order and minimality) can both be *proved* on the basis of von Neumann's definition of the ordinals.

These proofs depend upon a key principle of set theory (ontology of the multiple): the Axiom of Foundation.[4] This axiom says that every situation (every pure multiple) comprises at least one term (one element) that has 'nothing in common' with the situation, in the sense that nothing of that which composes the term (no element of the element) is presented in the situation (belongs to the original multiple).

8.12. Let's return to the example of my cat (**7.6**). It is an element of the set of living beings, and it is composed of cells that are in turn elements of this set, if one grants that they are living organisms. But if we decompose a cell into molecules, then into atoms, we eventually reach purely physical elements that don't belong to the set of living beings. There is a certain term (perhaps the cell, in fact) which belongs to the set of living beings, but none of whose elements belongs to the set of living beings, because those elements all involve only 'inert' physico-chemical materiality. Of this term, which belongs to the set but none of whose elements belongs to it, we can say that it grounds the set, or that it is a fundamental term of the set. 'Fundamental' meaning that on one side of the term, we break through that which it constitutes; we leave the original set, we exceed its presentative capacity.

8.13. Once more, let's leave living beings, cats, cells and atoms behind. Consider the singleton of the singleton of the void, that is, the set whose unique element is the singleton of the void, and which is written as ((0)). The element (0) of this set has as its only element the void, 0. Now the void *is not* an element of the original set ((0)), whose only element is (0), because the void 0 and the singleton of the void (0) are different sets. So (0) represents, in ((0)), a local foundation-point: it has no element in common with the original set ((0)). That which it presents qua multiple – that is, 0 – is not presented by ((0)), in the presentation in which it figures.

The Axiom of Foundation tells us that this situation is a law of being: every multiple is founded, every multiple comprises at least one element which presents nothing that the multiple itself presents.

8.14. The Axiom of Foundation has a remarkable consequence, which is that *no set can belong to itself*, that no multiple figures in its own presentation, that no multiple counts itself as one. In this sense, *being knows nothing of reflection*.

Take a set E which is an element of itself: E ∈ E. Consider the singleton of this set, (E). The only element of this singleton is E. So E must found (E). But this is impossible, since E belongs to E, and thus has in common with (E) that element which is itself. Since the axiom of foundation is a law of being, we must reject the original hypothesis: there does not exist any set that is an element of itself.

8.15. Returning to the crucial properties of the ordinals: They can be proved, once the axiom of foundation is assumed. I will do so here for the principle of minimality. For the principle of total order through belonging, see the note.[5]

Take an ordinal W_1 which possesses property P. If it is minimal, all is well. Suppose that it is not. In that case, there exist ordinals smaller than W_1 (and which therefore belong to W_1, since the order in question is belonging) and which also possess the property. Consider the set E of these ordinals ('gathering together' all those which possess property P and belong to W_1). Set E obeys the Axiom of Foundation. So there is an element W_2 of E which is an ordinal (since E is a set of ordinals) that possesses property P (since all the elements of E possess it) and that has no element in common with E.

But, since W_1 is an ordinal, it is transitive. So W_2, which belongs to it, is also a part of it: the elements of W_2 are all elements of W_1. If an element of W_2 possesses property P, then, since it is an element of W_1, *it must belong to E* (since E is the set of *all* the elements of W_1 possessing property P). Which cannot be, because W_2 founds E and therefore has no element in common with E. Consequently, no element of W_2 has the property P, and W_2 is minimal for this property. QED.

8.16. Thus is knitted the ontological fabric from which the numbers will be cut out.[6] Homogenous, intricate, rooted in the void, locally minimisable for every property, it is very much what we could call a *horizonal structure*.

9

Succession and Limit. The Infinite

9.1. In chapter 6, when we spoke of Dedekind's and Cantor's approaches to the notion of the ordinal (on the basis of well-orderedness), we saw that the whole problem was that after one ordinal comes another, well-determined, and that this series can be pursued indefinitely. We also saw that it was not at all the same thing to 'pass' from n to $n + 1$ (its successor) as to pass from 'all' the natural numbers to their beyond, which is the infinite ordinal ω. In the latter case, there is manifestly a shift, the punctuation of a 'passage to the limit'.

In the ontologised concept of the ordinals which von Neumann proposed and to which we dedicated chapter 8, do we once more find this dialectic between simple succession and the 'leap' to the infinite? And, more generally, how does the thorny issue of the existence of an infinite multiple present itself in this new context?

9.2. Let's apply ourselves firstly to the concept of succession.

We must take care here. The image of succession, of 'passage' to the next, is so vividly present in the immediate representation of number that it is often thought to be *constitutive* of its essence. I reproached J. A. Miller (see **3.17**) precisely for reducing the problem of number to the determination of that which insists in its successional engenderment. I held that the law of the serial passage across the numeric domain, a law which is imposed on *us*, does not coincide with the ontological immanence of number as singular form of the multiple.

Consequently, if we find the idea of succession once again in von Neumann's conception of the ordinals, it too must yield to the process of ontologisation. Our goal will be to discover, not so much a principle of passage as an intrinsic qualification of *that which* succeeds, as opposed to that which does not. What counts for us is not succession, but the *being of the successor*. The repetitive monotony of Peano's +1 does not concern us any longer: what we want to think is the proper being of that which can only be attained in the modality of the additional step.

9.3. Let's consider an ordinal W, in von Neumann's sense (a transitive set all of whose elements are transitive).

A set, then, whose elements are:

- all of the elements of W;
- W itself.

So, to everything that composes the multiple W, we 'add' *one* supplementary element, namely W itself. And it is indeed a question of the adjunction of a *new* element, since we know (it is a consequence of the axiom of foundation, compare **8.14**) that W is never an element of itself.

A non-operational form of +1 can be seen emerging here: it is not a matter of an extrinsic addition, of an external 'plus', but of a sort of immanent torsion, which 'completes' the interior multiple of W with the count-for-one of that multiple, a count whose name is precisely W. The +1 consists here in extending the rule of the assembly of sets to what had heretofore been the principle of this assembly, that is, the unification of the set W, which is thereafter aligned with its own elements, counting *along with them*.

9.4. An example of the procedure. We have demonstrated that set D, which is written (0,(0)), and which is the pair of the void and the singleton of the void, is an ordinal (it is transitive and all its elements are transitive). Our non-operational definition of +1 consists in forming the set of the *three* following elements: the two elements of D and D itself. We write this as (0,(0),(0,(0))) (the 'whole' D is found in the third position). Call this triplet T. We can now demonstrate that:

- T is transitive. Its first element, 0, is a universal part, and so it must be a part of T; its second element, (0), is the singleton of its first element, 0. So it is also a part of T (see **7.11**). Its third element

$(0,(0))$ is nothing but the 'gathering together', the forming into a pair, of these first two. So it is also a part. So every element of T is a part, and T is transitive.

- All the elements of T are transitive. Given that we have shown that D is an ordinal, we have duly shown that its elements, 0 and (0), are also transitive. We have equally demonstrated that it itself, $(0,(0))$, is transitive. And these are precisely the three elements of T.

So T, obtained by 'adjoining' D to the elements of D, is a von Neumann ordinal: a transitive set all of whose elements are transitive.

9.5. The reasoning we just followed can easily be generalised. For any ordinal W whatsoever, everything will follow just as for T: the set obtained in adjoining W itself, *as* an element, to W's elements is an ordinal.

We 'step' from W to a new ordinal by adjoining to W's elements a single additional element (this, now, allows us to lift a corner of the veil on the identity of our example T: just as D was two – I would like to say *the being of number Two* – T is none other than the number Three).

The fact that one steps from W to a new ordinal, whose elements are those of W supplemented by the one-name of their assembly, by way of a sort of immanent +1, justifies the following definition: *we will call the ordinal obtained by joining W to the elements of W, the* successor *of the ordinal W, and will denote it by S(W).*

So, in our example, T (three) is the successor of D (two).

9.6. The idea of the 'passage' from two to three, or from W to S(W), is, in truth, purely metaphorical. In fact, from the start *there are* figures of a multiple–being, D and T, and what we have defined is a *relation* whose sole purpose is to facilitate *for us* the intelligible passage through their existences. Finitude demands the binding of the un-binding of being. We therefore think, in the succession T = S(D), a relation whose basis is, in truth, immanent: T has the structural property, verifiable in its ontological composition, of being the successor of D, and it is no more than a necessary illusion to represent T as being constructed or defined by the relation S, which connects it externally to D.

A more rigorous philosophical approach consists in examining the ordinals in themselves and in asking ourselves whether they possess the *property of succeeding*. For example, T possesses the property of succeeding D, *recognisable in itself from the fact that D is an element of T*, and, what's more – as we shall see – that D is an element that can be immanently distinguished (it is 'maximal' in T).

We will call *successor ordinal* an ordinal that possesses the property of succeeding.

So T is a successor ordinal.

9.7. It might be objected that the property 'succeeds W' is still latent in the intrinsic concept of successor, and therefore that we have failed to establish ourselves in the ontological unbinding. This objection can be alleviated.

Let's consider an ordinal W having the following, purely immanent, property: amongst the elements of W, there is *one* element, say w_1, of which *all the other* elements of W are elements: if w_2 is an element of W different from w_1, then $w_2 \in w_1$. I say that W is necessarily a successor ordinal (in fact, it succeeds w_1).

For if this situation obtains, it is because W's elements are:

– on the one hand the element w_1;
– on the other, elements which, like w_2, are elements of w_1.

But, in reality, *all* the elements of w_1 are elements of W. For we know that belonging, $\in$, is a total order over the ordinals (see **8.10**). Now, all the elements of an ordinal are ordinals (**8.5**); specifically, all the elements of W are ordinals. w_1 is therefore an ordinal, and it follows that the elements of w_1 are all ordinals. These elements are connected to ordinal w_1 and W by the relation of total order that is belonging: if $w \in w_1$, since $w_1 \in W$, then $w \in W$ (transitivity of the order-relation).

Thus W is composed of all the elements of w_1, and w_1 itself: W is by definition the successor of w_1.

Let's agree to call the *maximal element* of an ordinal the element of that ordinal which is like w_1 for W: all the other elements of the ordinal belong to the maximal element. The reasoning above now permits us to make the following definition: *An ordinal will be called a successor if it possesses a maximal element.*

Here we are in possession of a totally intrinsic definition of the successor ordinal. The singular existence of an 'internal' maximum, located solely through the examination of the multiple structure of the ordinal, of the fabric of elementary belonging at its heart, allows us to decide whether it is a successor or not.

9.8. Since we now have an immanent, non-relational and non-serial concept of 'what a successor is', we can pose the question: Are there ordinals that *are not* successors?

9.9. The empty set, 0, is an ordinal that is not a successor. It obviously cannot succeed anything, since it has no elements and, to succeed, it must have at least one element, namely the ordinal that it succeeds.

Or, staying closer to the immanent characterisation: to be a successor, 0 must have a maximal element. Having no elements, it cannot be a successor.

Once again, we discover the void's function as ontological anchor: purely decided in its being, it is not *inferable* and, in particular, it cannot succeed: the void is itself *on the edge of the void*, there is no way it could follow from being, of which it is the original point.

9.10. All the ordinals that we have used in our examples, apart from the void, are successors. Thus (0) (which is the number 1) is the successor of 0. The number 2, whose being is (0,(0)), and which is composed of the void and 1, is the successor of 1. And our T (the number 3), which is composed of the void, 1, and 2 and is written (0,(0),(0,(0))), is the successor of 2. It is clear that we can continue, and will thereby obtain 4, 5, and, finally, any of the natural whole numbers, *all of which are successor ordinals*.

9.11. Does this mean that we have at our disposal a thinking of natural whole number? Not yet. We can say that 1, then 2, then 3, etc., if we think each in its multiple–being, are natural whole numbers. But, without being able to determine the *place* of their deployment, it is impossible for us to pass beyond this case-by-case designation and to propose a general concept of whole number. As Dedekind perceived, such a concept necessitates a detour through the infinite, since it is within the infinite that the finite insists. The only thing that we can say with certainty is that whole numbers are successor ordinals. But this is certainly not a sufficient characterisation of them: there might well be other successors that are not whole numbers, perhaps successors that are not even finite sets.

9.12. The question becomes: are there any other non-successor ordinals *apart from the void*?

Let's call these non-successor ordinals different from 0 (without yet knowing whether they exist) *limit ordinals*. We ask once more: do limit ordinals exist?

We are not yet in a position to decide upon this question. But we can prove that, *if they do*, they are structurally very different from successor ordinals.

9.13. No ordinal can come in between an ordinal W and its successor S(W). By this we mean that, given that the order-relation between ordinals is that of belonging, no ordinal W_1 exists such that we have the sequence $W \in W_1 \in S(W)$.

We know in fact that W is the maximal element of S(W) (see **9.7**). Consequently, *every* element of S(W) that is different from W belongs to W. Now, our supposed W_1 belongs to S(W). Therefore one of two things must apply:

- either W_1 is identical to W. But this is impossible, because we have supposed that $W \in W_1$, which would give us $W \in W$. But we know (**8.14**) that no set can be an element of itself;
- or W_1 is an element of W. But then it would not be possible that $W \in W_1$, since $W_1 \in W$.

It can be seen that ordinal succession is the schema of the 'one more step', understood as that which hollows out a void between the initial state and the final state. Between the ordinal W and its successor S(W), there is *nothing*. Meaning: nothing natural, no ordinal. We could also say that a successor ordinal delimits, just 'behind' itself, a gap where nothing can be established. In this sense, rather than succeeding, *a successor ordinal begins*: it has no attachment, no continuity, with that which precedes it. The successor ordinal opens up for thought a beginning in being.

9.14. A limit ordinal, if such a thing exists, is a different case altogether. The definition of such an ordinal is, please note, purely negative: it *is not* a successor; that is all that we know of it for the moment. We can also say: it does not possess a maximal element. But the consequences of this lack are considerable.

Take L, a supposed limit ordinal, and w_1, an element of this ordinal. Since w_1 is not maximal, there certainly exists an element w_2 of L which is larger than it: so we have the chain: $w_1 \in w_2 \in L$. But, since in its turn w_2 is not maximal, there exists a w_3 such that $w_1 \in w_2 \in w_3 \in L$. And so on.

Thus, when an ordinal belongs to a limit ordinal, a third party is intercalated into the relation of belonging, and, as this process has no stopping point, as there is no maximal element, it can be said that, between any element w of a limit ordinal L and L itself, there is always an 'infinity', in the intuitive sense, of intermediate ordinals. So it is in a strong sense that the limit ordinal does not succeed. No ordinal is the last to belong to it, the 'closest' to it. A limit ordinal is always equally 'far' from all the ordinals that belong to it. Between the element w of L and L, there is an infinite distance where intermediaries swarm.

The result is that, contrary to what is the case for a successor ordinal, a limit ordinal does not hollow out any empty space behind itself. No matter how 'close' to L you imagine an element w to be, the space between w and L is infinitely populated with ordinals. The limit ordinal L is therefore in a relation of *adherence* to that which precedes it; an infinity of ordinals 'cements' it in place, stops up every possible gap.

If the successor ordinal is the ontological and natural schema of radical beginning, the limit ordinal is that of the *insensible result*, of transformation without gaps, of infinite continuity. Which is to say that every action, every will, is placed either under the sign of the successor, or under the sign of the limit. Nature here furnishes us with the ontological substructure of the old problem of revolution (*tabula rasa*, empty space) and of reform (insensible, consensual and painless gradations).

9.15. There is another way to indicate the difference between successors and limits (which are for us the predicates of natural multiple–being).

The *union of a set E* is the set constituted by the *elements of the elements of E*. This is related to a very important operator of the ontology of the multiple, the operator of *dissemination*. The union of E 'breaks open' the elements of E and collects all the products of this breaking-open, all the elements contained in the elements whose counting-for-one E assures.

An example: take our canonical example of three, the set T that makes a triplet of the void, the singleton of the void and the pair of the void and its singleton. It is written $(0,(0),(0,(0)))$. What is the union of T?

The first element of T is 0, which has no elements. It therefore donates no elements to the union. The second element is (0), whose single element is 0. This latter element will feature in the union. Finally the third element is $(0,(0))$, whose two elements are 0 (which we already have) and (0). So in the end the union of T, the set of the elements of its elements, is composed of 0 and (0): it is the pair $(0,(0))$. That is to say, our D, or the number two. The dissemination of three is no other than two. We state in passing (this will be clarified in **9.18**) that the union of T is 'smaller' than T itself.

9.16. The position of ordinals with regard to union is most peculiar. Given that an ordinal W is transitive, all its elements are also parts. And this means that *the elements of the elements of W, which are also the elements of the parts of W, are themselves elements of W.* In the union of an ordinal we find nothing but the elements of that ordinal. That is to say that the union of an ordinal *is a part of the ordinal.* If we denote the set 'union of E' by $\cup$E, then, for every ordinal, $\cup$W $\subset$ W.

This property is characteristically natural: the internal homogeneity of an ordinal is such that dissemination, breaking open that which it composes, never produces anything other than a part of itself. Dissemination, when it is applied to a natural multiple, delivers only a 'shard' of that multiple. Nature, stable and homogenous, can never 'escape' its proper constituents through dissemination. Or: in nature there is no non-natural *ground*.

9.17. That the union of an ordinal should be a part of that ordinal, or that the elements of its elements should be elements, brings us to the question: are they *all*? Do we ultimately find not even a 'partial' part (or *proper* part, compare **4.12**), but only the ordinal we began with? It could well be that *every* element can be found as element of an element, since the internal fabric of an ordinal is entirely intricated. In that case, $\cup$W = W. Not only would dissemination return only natural materials, but it would restore the initial totality. The dissemination of a natural set would be a *tautological* operation. Which is to say that it would be absolutely in vain: we could then conclude that *nature does not allow itself to be disseminated.*

9.18. This seductive thesis is verified *in the case of limit ordinals*, if such a case exists.

Take any element w_1 whatsoever of a limit ordinal L. We have shown (in **9.14**) that between w_1 and L necessarily comes an intercalated element w_2, in such a fashion that we always have (whatever the element w_1) the chain $w_1 \in w_2 \in$ L. But, in addition, when we disseminate L the element w_1 will be found again in the union, as an element of w_2. Consequently, *every* element of L features in $\cup$L, the union of L. And, as we have seen, conversely (**9.15**), that every element of $\cup$L is an element of L (since $\cup$L $\subset$ L), it only remains to conclude that the elements of L and those of $\cup$L are exactly the same. Which is to say that L is identical to $\cup$L.

To dissemination, the limit ordinal opposes its infinite self-coalescence. It is exemplarily natural, in so far as, in being 'dissected', its elements do not alter. It *is* its own dissemination.

9.19. A successor ordinal, on the other hand, resists being identified with its dissemination. It remains *in excess* of its union.

Let's consider a successor ordinal W. By definition it has a maximal element w_1. Now it is impossible that this element should be found in the union of W. If it were found, that would mean that it was the element of another element, w_2, of W: so $w_1 \in w_2$, and w_1 would not be maximal. The maximal element w_1 *necessarily makes the difference between W and* $\cup$W. There is at least one element of a successor ordinal that blocks the pure and simple disseminative restoration of its multiple–being. A successor, unlike a limit, is 'contracted', altered, by dissemination.

9.20. In my view, this contrast is of the greatest philosophical importance. The prevailing idea is that what happens 'at the limit' is more complex, and also more obscure, than that which is in play in a succession, or in a simple 'one more step'. For a long time philosophical speculation has fostered a sacralisation of the limit. What I have called elsewhere[1] the 'suture' of philosophy to the poem rests largely upon this sacralisation. The Heideggerian theme of the Open, of the deposition of a closure, is the modern form of the assumption of the limit as a wrenching away from counting, from technique, from the succession of discoveries, from the seriality of Reason. There is an *aura* of the limit, and an unbeing of succession. The 'heart come from another age' aspires (and this horizon-effect is only captured, so it seems, by the poem) to a movement across those 'infinite meadows where all time stands still'.[2]

What the ontology of the multiple (based in a contemporary Platonism) teaches us is, on the contrary, that the difficulty resides in succession, and that there, also, resides *resistance*. Every true test for thought originates in the localisable necessity of an additional step, of an unbroachable beginning, which is neither *fused* through the infinite replenishment of that which precedes it, nor identical to its dissemination. To understand and endure the test of the additional step, such is the true necessity of time. The limit is a recapitulation of that which composes it, its 'profundity' is fallacious; it is in virtue of its having no *gaps* that the limit ordinal, or any multiplicity 'at the limits', attracts the evocative and hollow power of such a 'profundity'. The empty space of the successor is more redoubtable, it is *truly* profound. There is nothing more to think in the limit than in that which precedes it. But in the successor there is a crossing. The audacity of thought is not to repeat 'to the limit' that which is already entirely retained within the situation which the limit limits; the audacity of thought consists in crossing a space where nothing is given. We must learn once more how to succeed.

9.21. Basically what is difficult in the limit is not what it gives us to think, but its *existence*. And what is difficult in succession is not its existence (as soon as the void is guaranteed, it follows ineluctably) but that which begins in thought with this existence.

And so, speaking of the limit ordinal, the question returns, ever more insistent: do limit ordinals exist? On condition of the existence of the void, there is 1, and 2, and 3 . . ., all successors. But a limit ordinal?

The reader will have realised: we find ourselves on the verge of the decision on the infinite. No hope of *proving* the existence of a single limit ordinal. We must make the great modern declaration: the infinite exists, and, what is more, it exists in a wholly banal sense, being neither revealed (religion), nor proved (mediaeval metaphysics), but being simply decided, under the injunction of being, in the form of number. All our preparations amount only to saying, to being able to say, that the infinite can be thought *in the form of number*. We know it, at least for that which falls within the natural ontological horizon of number: the ordinals. That is infinite which, not being void, meanwhile does not succeed. It is time to announce the following:

Axiom of Infinity. **A limit ordinal exists.**

10

Recurrence, or Induction

10.1. A momentary pause to begin with: let's recapitulate what the ordinals give us to think as regards being qua being, from the viewpoint of a philosophy informed by mathematical ontology.

10.2. The ordinals are, because of the internal stability of their multiple–being (the maximal identity between belonging and inclusion, between 'first' presentation through the multiple, as element, and re-presentation through inclusion, as part) and the total homogeneity of their internal composition (every element of an ordinal is an ordinal), the ontological schema of *natural* multiplicity.

10.3. The ordinals do not constitute a set: no multiple–form can totalise them. There exist pure natural multiples, but *Nature does not exist*. Or, in Lacanian terms: Nature is not-all, just as is being qua being, since no set of all sets exists either.

10.4. The anchoring of the ordinals in being as such is twofold.

The absolutely initial point that assures the chain of ordinals of its being is the empty set 0, decided axiomatically as secularised form, or number–form, of Nothingness. This form is nothing other than the situation–name of being qua being, the suture of every situation–being, and of every language, to their latent being. The empty set being an ordinal, and therefore a natural multiple, we might say: the point of being of every situation is natural. Materialism is founded upon this statement.

10.5. The point-limit that 'restarts' the existence of the ordinals beyond Greek number (the finite natural whole numbers; on Greek number, see chapter 1) is the first infinite set, ω, decided axiomatically as a secularised form – and thus entirely subtracted from the One – of infinite multiplicity.

From this point of view, the ordinals represent the *modern* scale of measurement (conforming to the two crucial decisions of modern thought) of natural multiplicity. They *say* that nothingness is a form of natural and numerable being, and that the infinite, far from being retained in the One of a God, is omnipresent in nature, and, beyond that, in every situation–being.

10.6. *Our* passage through the ordinals (or the limits of our representation of them) arranges them according to an untotalisable sequence. This sequence 'starts' with 0. It continues through the natural whole numbers $(1, 2, \ldots, n, n+1, \ldots,$ etc.), numbers whose form of being *is composed of the void* (in the forms $(0), (0,(0)), (0,(0), (0,(0))), \ldots,$ etc.). It is continued by an infinite (re)commencement, guaranteed by the axiom 'a limit ordinal exists', which authorises the inscription, beyond the sequence of natural whole numbers, of ω, the first infinite ordinal. This recommencement opens a new series of successions: $\omega, \omega + 1, \ldots, \omega + n, \ldots,$ etc. This series is closed beyond itself by a *second* limit ordinal, $\omega + \omega$, which inaugurates a new series of successions, and so on. Thus we achieve the representation of a series of ordinals, deployed with no conceivable stopping point, which transits within the infinite (beyond ω) just as in the finite.

10.7. The ordering principle of this sequence is in fact belonging itself: given two ordinals W_1 and W_2, then $W_1 \in W_2$, *or* $W_2 \in W_1$, *or* $W_1 = W_2$. Belonging, a unique ontological relation because it governs the thinking of multiple–being as such, is also that which totally orders the series of ordinals. So that, if W is an ordinal and S(W) its successor, then $W \in S(W)$. So that, if n is a natural whole number (a finite ordinal) and n' a 'larger' whole number, then $n \in n'$. And so that, for any natural whole number n whatsoever, $n \in \omega$ (the first infinite ordinal), etc.

10.8. There are three types of ordinal (after the modern decisions which *impose* the void and the infinite):

1   The empty set, 0, is the inaugural point of being.
2   The successor ordinals adjoin to their predecessor *one* element, namely that predecessor itself. The successor of W is called S(W).

W is the maximal element in S(W), and the presence of a maximal element allows us to characterise successors in a purely immanent (non-serial) fashion. Successor ordinals give us a numerical schema for what it means to say 'one more step'. This step consists always in supplementing all that one has at one's disposal, with a unique mark for that all. To take 'one more step' comes down to making one of all of the given multiplicity, and adjoining that one to it. The new situation is 'maximalised': it contains one term that dominates all the others.

3   The limit ordinals have no maximal internal element. They mark the beyond proper to a series without stopping point. They do not succeed *any* particular ordinal, but it can be said that they succeed *all* the ordinals of the sequence of which they are the limit. No ordinal in this sequence is 'closer' to the limit ordinal than any other. For a third ordinal, and ultimately an 'infinity' (in the intuitive sense of a series with no stopping point) of ordinals, will intercalate themselves (according to the order-relation, which is belonging) between every ordinal of the sequence and the limit ordinal. The limit ordinal adheres to everything that precedes it. This is specifically indicated by its identity with its own dissemination ($L = \cup L$). The limit *totalises* the sequence, but does not *distinguish* any particular ordinal within it.

10.9. Just as a limit ordinal is structurally different from a successor ordinal (as regards the internal maximum, and as regards dissemination), so the 'passage to the limit' is an operation of thought entirely different from 'taking one more step'.

Succession is, in general, a *more difficult* local operation than the global operation of passage to the limit. Succession gives us more to think about than does the limit. The widespread view to the contrary stems from the fact that, not being 'absolutely modern', we still tend to sacralise the infinite and the limit, which is to say: retain them still in the form of the One. A secularised thought, subtracted from the One and the sacred, recognises that the most redoubtable problems are local problems, problems of the type: 'How to succeed?', 'How to take one more step?'.

10.10. The space of the ordinals allows us to *define* the infinite and the finite. An ordinal is finite if, in the chain of order governed by belonging, it comes before ω. It is infinite if it comes after ω (including ω itself).

We will find that, just as Dedekind's intuition suggested, only the existence of an infinite ordinal permits us to define the finite. Modern thought says that the first situation, the banal situation, is the infinite. The finite is a secondary situation, very special, very singular, extremely rare. The obsession with 'finitude' is a remnant of the tyranny of the sacred. The 'death of God' does not deliver us to finitude, but to the omnipresent infinitude of situations, and, correlatively, to the infinity of the thinkable.

10.11. The final synthetic recapitulation of the fact that the ordinals give us to think being qua being, in its natural proposition, is complete. Now we must turn towards *our* capacity to traverse and to master rationally this donation of being. One way to do so is simply to proceed, in this boundless fabric, to the carving-out of Number.

10.12. It is a blessing for our subjective finitude that the authority – properly without measure – of natural multiplicities allows that diagonal of passage, or of judgement, which is reasoning by recurrence, also called complete induction and, in the case of infinite ordinals, 'transfinite induction'. In fact this alone allows us, in treating of an infinite domain (and even, if we consider the ordinals, one that is infinitely infinite), to anticipate the *moment of conclusion*.

Suppose that we wish to show that *all* ordinals possess a certain property P. Or that we wish to establish rationally, with a proof, a universal statement of the type: 'For all x, if x is an ordinal, then $P(x)$'. How can this be achieved? It is certainly impossible to confirm case by case that it is so: the task would be infinitely infinite. Neither is it possible to consider the 'set of ordinals', since such a set does not exist. The 'all of the ordinals', implied in the universal quantifier of the statement 'for all x', cannot be converted into 'all the elements belonging to the set of ordinals'. Such a set is inconsistent (see **6.11**). It is precisely the alleviation of this impasse that is the business of reasoning by recurrence.

10.13. Reasoning by recurrence combines *one* verification and the demonstration of *one* implication. Once in possession of these two moments, the structure proper to the ordinals authorises the universal conclusion.

Take property P. We begin by confirming that the empty set 0 possesses this property; we test P for the 'case' of 0. If the empty set does not possess the property P, it is pointless to pursue the investigation. If one ordinal, 0, does not have property P, it is certainly false that all ordinals do. Suppose, then, that the statement P(0) is true, that the test in the case of 0 is positive.

We will now try to prove the following implication: *if* all the ordinals that precede some ordinal W (according to the total ordering of the ordinals, which is belonging) have the property P, *then* W also has it.

Note that this implication does not tell us that an ordinal with property P *exists*. It remains in the hypothetical register, according to the general pattern: 'if x is so, then what follows x is so'. The implication is really universal, it does not specify any ordinal W. It says only that, for every ordinal W, supposing that those which precede it in the chain of ordinals satisfy P, one is compelled to admit that W satisfies it also.

It is usually necessary to divide this demonstration (supposing that it is possible, which obviously depends on property P), by treating the case where we suppose W to be a successor separately from the case where we suppose it to be a limit (since W is any ordinal whatsoever, it could be one or the other). Reasoning by recurrence, as we saw in the central implication that constitutes it, strongly binds that which is the case for an ordinal W to that which is that case for the ordinals that precede it. Now the relationship of a limit ordinal to the anterior ordinals (one of infinite adherence) differs radically from that of a successor (which, between itself and its predecessor, clears an empty space). Because of this, the procedures of thought and of proof put into play in the two cases are usually heterogeneous. And, as we might expect, given the philosophy of this heterogeneity (cf. **9.19**), it is generally the case of the successor that is the most difficult.

Assume that we have verified the truth of P(0), and that we have proved the implication 'if, for every ordinal w that precedes W (that belongs to W: order is belonging), it is the case that P(w), then it is also the case that P(W)'. We can conclude that *all* ordinals satisfy P, in spite of the fact that this 'all' not only alludes to an infinitely infinite immensity of multiples, but that, even so, it does not make an All. It is truly the infinite and inconsistency 'conquered word by word'.

10.14. What authorises such a passage to 'all', such an ambitious 'moment of conclusion'? The authorisation is granted us by a fundamental property of the ordinals as ontological schema of the natural multiple: their 'atomistic' character, the existence, for every property P, of a *minimal* support for this property as soon as one ordinal possesses it. (See **8.10** and **8.15**).

If the conclusion were false – if it were not the case that all ordinals possess property P – that would mean that there was at least one ordinal which did not possess property P. This ordinal would then possess the property not-P, not-P meaning simply 'not possessing property P, being a non-P'.

But, if there exists an ordinal that possesses property not-P, *there exists a smallest ordinal which possesses this property not-P*, by virtue of the atomistic principle, the principle of minimality. And, since it is the smallest to possess property not-P, all those which are smaller than it must possess property P.

We could object: these ordinals 'smaller than it' may not exist, because it is possible that the minimal ordinal for the property not-P is the void, which is not preceded by anything. But no: since (first moment of our procedure) we have verified precisely that 0 possesses the property P, the minimal ordinal for not-P cannot be 0. Thus it does make sense to speak of ordinals smaller than it; they exist, and must all possess property P.

Now our central implication, supposed proved, said exactly that, if all the ordinals smaller than a given ordinal possess property P, then that ordinal also possesses it. We have reached a formal contradiction: that the supposed minimal not-P must be a P. It is necessary then to conclude that this latter does not exist and that therefore all ordinals do possess property P.

Thus the ontological substructure of natural mutiplicities comes to found the legitimacy of recurrence. Our verification (the case of 0) and our demonstration (if $P(w)$ for all w such that $w \in W$, then $P(W)$ also), if it is possible (which depends on P . . . and on our mathematical know-how), authorises the conclusion for 'all ordinals'.

10.15. We have remarked, in studying Peano's axiomatic (see 5.3) that reasoning by recurrence is a fundamental given of *serial* numericality, of which the natural whole numbers are an example. It is quite natural that it should extend to that 'universal series' composed by the ordinals. But the great difference is that, whereas in Peano the principle of induction or recurrence is an axiomatic form or a formal disposition, here, since it is founded in being (in the theory of the pure multiple), it is a *theorem* – that is, a property *deducible* from the ordinals.

It is of the essence of the natural multiple, which escapes all totalising thought, to submit itself nonetheless to that intellectual 'capture' which is the inductive schema. Here, once more, being is found to be amenable to thought in that form of Number which is the conclusion for 'all', proceeding both from the verification for one only (here, 0) and from a general procedure which transfers the property of what comes 'before' (predecessor or endless series, depending on whether it is a case of a successor ordinal or a limit ordinal) to what comes 'after'. Number is that which accords being to thought, in spite of the irremediable excess of the former over the latter.

10.16. Reasoning by recurrence is a proof-procedure for universal statements concerning ordinals. It allows us to conclude. But there is a more important usage of recurrence, or of transfinite induction, one which allows us to *attain the concept*. This is *inductive definition*.

Suppose that the aim of our thinking is not to prove that this or that type of multiple, for example ordinals, has property P, but to define property P in a way that would allow us *then* to test it on multiples. A well-known difficulty in such a case is that *we don't know in advance whether a property defined in language is 'applicable' to a pure multiple without inconsistency resulting*. We have seen, for example (in **2.11**), that the property 'not being an element of itself' does not apply to any existing set, and that its perfect formal correctness does not alter the fact that, handled without care, it leads to the ruin, by way of inconsistency, of all formal thought. But how can we introduce limitations and guarantees, if language alone cannot support them? The procedure of definition by recurrence, or inductive definition, answers this question.

10.17. What will found the legitimacy of the procedure this time is the fact that, with the ordinals, we have at our disposal a sort of universal scale, which allows us to define property P *at successive levels*, without exposing ourselves to that danger of inconsistency that attends on any supposition of an All. Inductive definition is a *ramification* of the concept: property P will not be defined 'in general', but always as indexed to a certain level, and the operators of this indexation will be the ordinals. Here, once again, being comes to the aid of finitude, in assuring for our thought, which the domain of being as pure multiple exceeds on all sides, that it can proceed in steps, in fragments.

10.18. In conformity with the typology of ordinals, which distinguishes three types (the void, successors, limits), our procedure is divided into three.

1 We first define *explicitly*, with a statement, level 0 of the property. An explicit definition assumes that we have a property – say, Q – *already* defined, and that we can affirm that level 0 of P – say P_0 – *is equivalent* to Q. We would then have: $P_0(x) \leftrightarrow Q(x)$.

2   We then say that, *if* level w of P is defined, P_w, *then* level $S(w)$, that is, $P_{S(w)}$, is defined through an explicit procedure to be indicated. To say that P_w is defined is to say that there is a property – call it R – already defined such that P_w is equivalent to it, so $P_w(x) \leftrightarrow R(x)$. The existence of an explicit procedure enabling us to pass from the definition of P_w to that of $P_{S(w)}$ means that there is a function f that assures the passage of R (which defines P_w) to a property $f(R)$ which will define $P_{S(w)}$. Finally, we can say that 'x has the property $P_{S(w)}$' means 'x has the property $f(R)$', or that f, which permits the 'passage' from the definition of P_w to that of $P_{S(w)}$, is an explicit operation on R, fixed once and for all.

3   Finally, we will say that, *if* all the levels of P below a limit ordinal L have been defined, say: $P_0, P_1, \ldots, P_n, P_{n+1}, \ldots$, *then* level L of P, say for example P_w, is defined by a 'recollection' that can be explicated by that which defines all the levels anterior to it (in this process, union or dissemination generally plays a decisive role, for reasons given in **9.17**). Usually we have something like: for a given x, $P_L(x)$ is true, if there exists a level below L, call it w, where $w \in$ L, for which $P_w(x)$ is true. The limit level, in conformity with its essence, will assume all the inferior levels and will not introduce anything new.

Thus we will have at our disposal not just a single concept P, but an infinite and infinitely ramified family of concepts, from P_0, explicitly defined, up to the more considerable ordinal indexations P_w, passing through $P_n, P_\omega, P_{\omega+n}$, etc. We will then be able to say that concept P, as unique concept, is *defined by transfinite induction*, in the following sense: for a given x, $P(x)$ will be true if and only if there exists an ordinal W such that x possesses the property at level W. We would have the following equivalence: $P(x) \leftrightarrow$ 'there exists a W such that $P_W(x)$'.

So the inductive mastery of the concept passes by way of its ordinal ramification, and by way of the equivalence between 'the concept P holds for x' and 'the concept P holds for x *at level W of that concept*'. This equivalence *avoids all mentioning of the All*. It tests the property P not 'in general', but on *one* level, thus freeing it from paradoxes of inconsistency.

10.19. I shall give a most interesting example; its interest is both intrinsic (it sheds a keen light on the general structure of the theory of the pure multiple, or ontology: it proves that, thought in their being qua being, multiples are stratified) and methodological (we will see clearly the functioning of levels in the definition of the concept).

The underlying idea is to define, for each multiple, an ontological *rank*, indexed on the ordinals, which measures its 'distance', in a certain sense, from that initial suture which is the empty set. We could also say that the rank is a measure of the *complexity* of a set, of the immanent intrication of the instances of the void that constitute it.

Naturally, it is impossible to speak of 'all' sets: to do that it would be necessary to collect them as the elements of a set of all sets, which would be inconsistent. The prudent, gradual approach of the inductive procedure is indispensable here.

The two important operations of set theory which allow one to 'step' from one set to another are:

1   Union, or the set of elements of elements of the initial set; the operation of dissemination, which we have already met (compare **9.15**). Given a set E, we denote its union by $\cup$E.
2   The set of parts, which consists of 'gathering together' to make one all the *parts* of the initial set, all that is included in that set (on belonging and inclusion, see **7.3**). We denote by p(E) the set of the parts of E. Note that the *elements* of p(E) are the *parts* of E: if $e \in p$(E), then $e \subset$ E.

We will construct the hierarchy of ranks by means of these two operations. The property we will try to define through transfinite induction, according to the method explained in **10.18**, will be denoted by R(x), to be read as: 'x possesses a rank' (or: 'x is well-founded'). Our three steps will be as follows:

1   *Explicit definition of the property at level 0.* We propose that $R_0(x)$ *is not true for any* x, in other words that $R_0(x)$ is equivalent to $x \in 0$.
2   *Uniform treatment of successive levels.* We posit that $R_{S(w)}(x)$ is true if and only if x belongs to the set of parts of the set constituted by all the z which satisfy R_w. In other words, the rank at successor level S(w) is the set of parts of the rank defined for the level which the predecessor w indexes. This can be written as follows: $R_{S(w)}(x) \leftrightarrow ((y \in x) \rightarrow R_w(y))$: if x satisfies $R_{S(w)}$, the elements of x satisfy R_w, and consequently x is a part of the set of sets which satisfy R_w. We could also write, denoting by R_w the set of x for which $R_w(x)$ is true:

$$(x \in R_{S(w)}) \leftrightarrow (x \subset R_w) \leftrightarrow (x \in p(R_w))$$

3   *Uniform treatment of limit levels.* As would be expected, it is
    union that is at work here. We will say that $R_L(x)$ is true if x is
    of a rank whose index is smaller than L, that is, if there exists
    a $w \in L$ for which $R_w(x)$ is true. Thus the rank R_L recollects all
    the elements of the ranks below it; it is the union of these ranks.
    With the same conventions as above, we can write: $(x \in R_L)$
    $\leftrightarrow x \in \cup R_w$ for all w smaller than L.

Property R is thereby totally defined by induction. We will say that
x possesses a rank, or that $R(x)$ (without index) is true, if a (successor
or limit) ordinal w exists for which $R_w(x)$ is true. This property
'means' that one arrives at the complexity of x, beginning from 0
(which defines level R_0 of the property), through the successive
employment of union and of passage to the parts, an employment
whose 'length' is measured by an ordinal: the smallest ordinal w for
which $R_w(x)$ is true.

10.20. That this procedure really 'works', that it makes sense ultimately to speak of *the* property R, however, is not self-evident. The
generosity of natural being consists in the fact that one can *prove* the
effectivity of this ramified determination of the concept.[1]

Thus thought proceeds in its passage through being, under the
universally intricated and hierarchised rule of Nature, which doesn't
exist, but provides measurable steps. Number is accessible to us
through the law of such a passage, at the same time as it sets the
conditions – as we saw with the ordinals – for this passage itself.
Number is that through which being *organises* thought.

11

Natural Whole Numbers

11.1. The ordinals directly give us the Greek numbers: natural whole numbers. We are even in a position to attach a new, non-Greek, legitimacy to the adjective 'natural' which mathematicians, with the symptomatic subtlety of their nominations, adjoin to the civil status of these numbers: they are 'naturals' by virtue of the fact that, within the realm of the finite, they coincide purely and simply with the ordinals, which constitute the ontological schema of the pure natural multiple.

For it is 'natural' to identify, in its being, the *place of number* (that is, of whole number) – a place whose existence Dedekind vainly tried to secure on the basis of the consideration of 'all the possible objects of my thought' – with the first infinite ordinal ω, whose existence we decide, under the modern injunction of being, as we enounce the axiom 'a limit ordinal exists'.

11.2. To say that ω is the place of whole number has a precise set-theoretical meaning: what 'occupies' the place is that which belongs to it. Now, not only do all ordinals that precede a given ordinal belong to it; they constitute *all* the elements of that initial ordinal.

In fact, we know that total order over the ordinals is really belonging (see. **8.10**). And, consequently, an ordinal smaller than a given ordinal W is precisely an ordinal that belongs to W. The image of an ordinal (for example, one larger than ω) is as follows:

$$0 \in 1 \in 2 \in \ldots \in n \in n+1 \in \ldots \in \omega \in \omega+1 \in \ldots \in W$$

where all the numbers in the chain of belonging constitute precisely the elements of W. Visualised like this, the ordinal W appears as a sequence of 'embedded' ordinals, whose 'length' is exactly W. There are W links in the chain in order to arrive at W. We might also see an ordinal W, containing exactly W ordinals (all those that precede it), as *the number of that of which it is the name*. Which is another way of saying that it is identified with the place where its predecessors insist, being the recollection of that insistence.

Thus the definition of natural whole numbers is entirely limpid: an ordinal is a natural whole number if it is an element of the first limit ordinal ω. In which case, the structure of the place of number is:

$$0 \in 1 \in 2 \in \ldots \in n \in n \in n+1 \in \ldots \in \omega$$

But we must take care to note that ω itself, which is the name of the place, *is not a part of it*, since no set belongs to itself (cf. **8.14**). The place of whole number, ω, is not an element of that place, *it is not a whole number*. As ω is the *first* limit ordinal, it follows that all whole numbers, except the empty set 0 of course, are successors.

11.3. An attentive reader might object as follows: I say that ω is the first limit ordinal. But am I sure that a 'first' limit ordinal exists? The Axiom of Infinity (**9.20**) says only: '*a* limit ordinal exists', it does not specify that this ordinal is 'the first'. What authorises us to call ω the 'first limit ordinal', or first infinite ordinal? It could well be that, as soon as I announce that 'a limit ordinal exists', a multitude of them appear, none of which is 'first'. There could be an infinite descending chain of such ordinals, just like the descending chain of negative numbers which, it is clear, has no first term: no whole negative number is 'the smallest', just as no whole positive number is 'the largest' (this second point in fact comes back to saying that ω, the beyond and the place of the series of positive numbers, is a limit ordinal).

But if I cannot unequivocally determine and fix the first limit ordinal, then what becomes of my definition of whole numbers?

11.4. We can overcome this objection, once more, thanks to that great principle of natural multiples that is minimality. We know that, given a property P, *if* an ordinal exists that satisfies that property, *then* there is one and only one minimal ordinal that satisfies it. Take the property 'being a limit ordinal'. There certainly exists an ordinal that satisfies it, since the Axiom of the Infinite says precisely that.

Thus, there exists one and only one limit ordinal that is minimal for this property. Consequently we can speak without hesitation of a 'first limit ordinal', or of the 'smallest limit ordinal', and it is to this unique ordinal that we give the proper name ω. There is therefore no ambiguity in our definition of natural whole numbers.

11.5. We must never lose sight of the fact that notations of the type 1,2,*n*, etc. are *ciphers*, in the sense of codes, which serve to designate multiples fabricated from the void alone. We have known for a long time (already in **8.3**) that 1 is in reality the singleton of the void, that is, (0), that two is the pair of the void and the singleton of the void, that is, (0,(0)), that three is the triplet of the void, the singleton of the void, and the pair of the void and singleton of the void, that is, (0,(0),(0,(0))), etc. To exhibit further this weaving of the void with itself, let's also write down the real being of the cipher 4: (0,(0),(0,(0)),(0,(0),(0,(0)))).

Evidently 4 is a set of four elements, in the order 0, then (0), then (0,(0)), then (0,(0),(0,(0))). These four elements are none other than zero, 1, 2 and 3. The elements of a whole number comprise precisely all those numbers that precede it, which is not surprising, since we have shown above that this is the innermost structure of every ordinal (**11.2**). We could write: 4 = (0,1,2,3). And, as we have said, to pass from 3 to 4 (as from any *n* to *n* + 1), we 'adjoin' to the elements of 3 (or of *n*) the number 3 itself (or the number *n*). Which is not surprising, since this is the general definition of succession in the ordinals (**9.6**).

It would obviously be impossible to use the procedure of succession to 'step' from some whole *n*, no matter how large, to the first limit ordinal ω. This is because ω, let us repeat, is not a whole number, it is the place of such numbers. An important law of thought emerges here (one which, we might say in passing, the Hegelian figure of Absolute Knowledge, supposed to be the 'last' figure of Consciousness, contravenes), which states that *the place of succession does not itself succeed.*

11.6. Once we have at our disposal the place of natural whole numbers, their multiple–being which weaves the void through the finite, and the law of succession as law of *our* passage through these numbers, we 'rediscover' the classical operations (addition and multiplication for example) through simple technical manipulations that arise from the general principles of inductive definition, or definition by recurrence, explained and legitimated on the basis of natural being in chapter 10. It is time to give a new example.

11.7. Take a given number, say for example 4. We want to define through induction a function F whose outcome will be as follows: for any number n whatsoever (therefore for *every* whole number, and there is an infinity of them), F(n) is equal to the sum $4 + n$. To achieve this, we have at our disposal only one operator: ordinal succession, since the only thing we know is that all the whole numbers except 0 are successors. We will proceed exactly according to the schema explained in **10.18**, except that we will not have to worry ourselves about the case of limit ordinals (since there are none before ω). We will, as before, use S(n) to denote the successor ordinal of the whole number n.

1   We will first state: F(0) = 4 (an explicitly given value, the underlying intuition being that $4 + 0 = 4$).
2   Then we will proceed to the successional induction by positing: F(S(n)) = S(F(n)). A regular and uniform relation between the value of the function for S(n) and its value for n, a relation that uses only what we already know; the operation of succession, defined in general on the ordinals. The underlying intuition is that $4 + (n + 1) = (4 + n) + 1$, to return to the usual 'calculating' notation where the successor of n is denoted by $n + 1$.

The value of the function is defined entirely by these two equations. Say, for example, that I wanted to calculate F(2). I would have the following mechanical sequence:

$$F(0) = 4$$

$$F(1) = (F(S(0)) = (S(F(0)) = S(4) = 5$$

$$F(2) = (F(S(1)) = (S(F(1)) = S(5) = 6$$

We can see clearly that such a schema is a true *definition of addition*, through the use of recurrence, on the basis of the operation of succession alone. Once we have obtained this general inductive schema of addition, multiplication can be similarly defined. Take the function to be defined, P(n), whose value is n multiplied by 4. We begin the induction this time with 1 and not with 0, stating that if F(n) is as above (defining $4 + n$ inductively):

$$P(1) = 4 \text{ (guiding intuition : } 4 \times 1 = 4)$$

$$P(S(n)) = F(P(n)) \text{ (guiding intuition : } 4 \times (n + 1) = 4 + (4 \times n))$$

These technical manoeuvres are of no direct interest. They serve only to convince us that whole numbers thought in their being (ordinals that precede ω, fabricated from finite combinations of the void) are indeed *also* the same ones with which we count and recount without respite, as the epoch prescribes us to do.

11.8. The philosophico-mathematical reconstruction of whole numbers is now complete. They do not derive from the concept (Frege), nor can their place be inferred from our possible thoughts (Dedekind), nor is their law limited to that of an arbitrarily axiomatised operational field (Peano). They are, rather, in the retroaction of a decision on the infinite, that part of number which being provides to us in its *natural* and *finite* figure.

The whole numbers are Nature itself, in so far as it is exposed to thought only to the limited extent of its *capacity for finitude*. Again, this exposition is possible only on condition of a point of infinity, the limit ordinal ω, the existential guarantee of whole number. This point of infinity is immense in relation to the whole numbers, since, subtracted from successoral repetition, it constitutes the place of their total exercise, a place without internal limits (succession can *always* continue). Nevertheless, it is minute in relation to the profusion of natural infinite being beyond its first term ω. Whole number is the form of being of the finite 'almost nothing' deployed by being qua being between the void and the first infinity.

11.9. It is only in an anticipation without solid foundation, and in homage to their antiquity, that we call the natural whole numbers 'numbers'. We have already remarked (8.8) that, still without a general concept of number at our disposal, it would be illegitimate to say that the ordinals were numbers. Now, the whole numbers are none other than the ordinals. And number, or rather Number, qualifies a type of being of the pure multiple which exceeds the ordinals. Until we have made sense of this type, in such a way that it becomes applicable to all species of number (whole, relative, rational, real, ordinal, cardinal), we can only speak of 'number' in a sense still insufficiently liberated from its operational intuition, or from the historical heredity of this signifier.

But our preparations are complete. The homage paid to the Greek numbers is only the last act of a vast introduction, genealogical and then conceptual. Now it is necessary to *define* Number.

3

Ontology of Number:
Definition, Order, Cuts, Types

12

The Concept of Number: An Evental Nomination

12.1. The first part of this book was historical and critical (a study of the great enterprises of the past). The second was constructive and conceptual (the determination of the ordinals as schema of natural multiplicity, on the basis of the concept of transitive sets). In this third part, we are going to proceed regressively, axiomatically: we shall begin with a general *definition* of Number, a remarkably simple definition involving only the concept 'ordinal'. Then, by way of increasingly specific determinations, we shall address the essential *attributes* of the resulting concept of Number: total order, the process of cutting, and finally – in the last place only – operations. In so doing, we shall demonstrate how all of our traditional numbers (the wholes, the rationals, the reals, and the ordinals themselves, conceived and handled as Numbers) are only *particular cases* of the general concept.

In my view, the three most important aspects of these proceedings are as follows:

1 Considerations of order and operations arise from the intrinsic, or ontological, definition of Number. Number is therefore not itself an operational concept, it is a particular figure of the pure multiple, which can be thought in a structural and immanent fashion. The operational dimensions are only subsequent traits. Number is not constructed; on the contrary, its very being makes possible all of the constructions in which we engage it.

2   The ordinals constitute the base material for the definition of Number, its natural ontological horizon. But, taken in all their generality, Numbers are 'non-natural' deductions from this natural material.

3   Our traditional numbers are only very specific cases, which certainly fall under the general and unified concept of Number but by no means exhaust it. There remains an innumerable immensity of Numbers we have not yet thought or used.

12.2. *Definition*: **A Number is the conjoint givenness of an ordinal and a part of that ordinal.**

A Number will be denoted by the letter N, followed by indices to distinguish between several different Numbers.

In other words, a number N is constituted by:

–   an ordinal W;
–   a subset F included in this ordinal, such that $F \subset W$.

The ordinal will be called the *matter* of Number, which we will denote by M(N).

The part of the ordinal will be called the *form* of the Number, which we will denote by F(N).

That part of the matter which is not in the form, that is, those elements of the ordinal W which are not in the part F(N), constitute the *residue* of the Number. We denote this by R(N). The residue is equal to the matter minus the form, and therefore to the set M(N) – F(N).

It is clear that, if we add together the form and the residue, we end up with the whole of the matter. Therefore, using $\cup$ to stand for union (see **9.15**): $F(N) \cup R(N) = M(N)$.

Since a Number is entirely determined by its matter (an ordinal) and its form (a part of that ordinal), it will often be convenient to write it as a pair (M(N), F(N)), with the convention that the ordinal–matter is written to the left, and the form to the right.

12.3. In examining this definition, the reader must observe a number of precautions.

(1) We are dealing with a pure definition, a priori for the moment: it is of no use, nor is it possible, to try to 'recognise' straightaway, in this definition, any of our familiar numbers.

I will give an example: take as matter the ordinal 1 (whose ontological composition is (0), the singleton of the void), and, as form, the ordinal 0, the void, which is of course a part of 1, as it is a part of every set (**7.9**). Using the above convention, we have the Number $N = (1,0)$. All we know is that, according to the definition (the result of an ordinal and a part of that ordinal), N *is a Number*. The signs '1' and '0' do not directly refer to any Number, since we have not yet even established that we are dealing with Numbers. In fact, what these signs 1 and 0 are going to indicate here – each on its own account, a matter and a form – *cannot be understood as Numbers*, since we do not discern in their writing the fundamental *dual* givenness of all Number: a matter *and* a form. A Number *must* involve two marks, that of its matter and that of its form: now '1' is only one mark, as is '0'. It would therefore be illusory to 'recognise' in the Number (1,0) any familiar number whatsoever, on the pretext that one 'recognises' 1 or 0. At the moment, we have in (1,0) only an abstract example of a Number, conforming to the concept of Number given in the definition.

Just as the prisoners in Plato's cave once again make the descent from the Idea (of Number) back down to the empirical (numbers), we will demonstrate *much later*, in chapter 16, that the Number (1,0) is the true concept of the familiar negative number −1. But at this stage it is essential that the reader consider the examples as simple clarifications of the definition of Number, and not seek to reconnect them to the cavernous empirical domain of numbers.

(2)   The matter of a Number is an ordinal; we have said enough about ordinals for there to be no mystery about this. On the other hand, the form of a particular Number is constrained only to be a *part* of that ordinal, a set included in the ordinal. The general concept of a part (or subset) is somewhat indeterminate and, when the matter happens to be infinite, offers no foothold for intuition. In particular, note:

– that this part might be empty (compare the example above);
– that this part might be the entire ordinal; if we take for matter the limit ordinal ω, and for form this same ordinal (which is a 'total part' of itself), we obtain a wholly permissible Number (conforming to the definition), which is written $N = (\omega, \omega)$; we will see in chapter 16 that there are excellent reasons to allow that this Number is none other than the ordinal ω itself, but at the moment this is not at all obvious;
– that this part does not necessarily have to be contained or connected as one; it could be dispersed, lacunary, composed of scattered elements, and so on; for example, if we take as matter the

limit ordinal ω, we can take as form the set constituted by the whole numbers 3, 587 and 1165. These three finite ordinals are all elements of ω, and therefore, taken together, they form a part of ω. We will have a quite permissible number N = (ω,(3,587,1165)), whose form has three completely separate elements.

12.4. These formal possibilities make a visualisation of Number difficult. We can imagine spatial designs somehow like this:

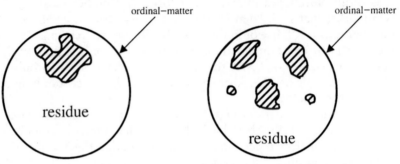

1) Number whose form is connected   2) Number whose form is dispersed

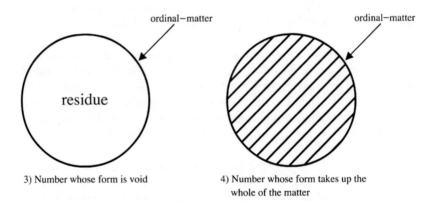

3) Number whose form is void   4) Number whose form takes up the whole of the matter

But doubtless the simplest way is to have recourse to a linear arrangement (see below). This figuration is based on ordinal linearity, conceived as a universal series from which the being of Number is deducted.

A line segment, whose supposed origin is the ordinal 0, represents the 'ordinal axis'. We mark with an asterisk * upon this axis the matter of the Number, an ordinal W. We mark with an emboldening of the line the form of the Number, part of its matter. The rest, left unchanged, represents the residue. If we want to represent a particular ordinal, we can do this with a little circle on the arrow, with the name above or below. With these conventions, a Number will look like this:

Once again, this type of drawing can aid comprehension, but can also be an encumbrance. Its principal failing, which it shares with the famous 'Venn diagrams' used to teach schoolchildren operations on sets (union, intersection, etc.), is that it habituates one to imagining that a part of a set is a sort of continuous whole, a compact neighbourhood. Now the sole prerequisite of a part is that it should contain only elements of the set of which it is a part. These elements might very well be highly dispersed, scattered to the far regions of the initial set, and the visual schema of a part, to indicate this dispersion, must be able to be punctured, fragmented, dismembered. The unfortunate thing is that the drawing then loses any intuitive value it might have had: one simply gets the impression that there are *many* parts. In looking at my lines and their emboldenings, one must always keep in view, conceptually, that there is no reason for the form of a Number to be a continuous segment, but that it could well be dispersed throughout the full extent of the ordinal–matter, as could the residue.

For example, the Number mentioned above, which has for matter the limit ordinal ω and for form the triplet (3,587,1165), must be represented somehow like this (with the additional complication that the infinity of ω is not truly 'commensurable' in a drawing):

12.5 The following section is entirely dedicated to a philosophical elucidation of our definition.

We will begin with the capital N with which I furnish Number.

In all attempts undertaken to determine the concept of number, the problems of terminology bring the weight of the event to bear upon the researcher.

Take for example the appellation 'irrational numbers'. It is truly astonishing to find such a designation at the heart of mathematical rationality. The doctrine of 'cuts' forged by Dedekind is nothing other than the determination – wholly rational and demonstrative – of the concept of irrational number. But exactly the same could be said for the theory of proportions in Euclid's *Elements*. It is clear, then, that 'irrational', in these mathematical texts whose rationality is transparent, paradigmatic even, no longer has any *signification*.

We might say that what makes itself known here is a symptom of the radical difference *between nomination and signification*. A signification is always distributed through the language of a situation, the language of established and transmitted knowledges. A nomination, on the other hand, emerges from the very inability of signification to *fix* an event, to decide upon its occurrence, at the moment when this event – which supplements the situation with an incalculable hazard – is on the edge of its disappearance. A nomination is a 'poetic' invention, a new signifier, which affixes to language that for which nothing can prepare it. A nomination, once the event that sustains it is gone forever, remains, in the void of significations.

Now, at the moment of the great Greek crisis of number, when the arrival of that at once inevitable and enigmatic event made it known that certain relationships (those, for example, of the diagonal of a square and its side) cannot be 'numbered' within the code of existing numbers, the word *alogos* arrived, saturating and exceeding the mathematical situation. This word designates that which, having no *logos*, nonetheless must be decided as number. It inscribes in a new situation of thought a nomination without signification: that of a number which is not a number.

Since that time, the word has lodged itself, without alteration, in mathematical language. It traverses translations, negligible but subsistent. Our word 'irrational' is unmindful of the import of the nomination *alogos* to the same extent that the word 'rational' retains little of the Greek *logos*. And, above all, this nomination has ended up taking on a univocal signification. But the contrast remains, and one can reactivate it – as I do – in between signification and that which, in the word that imparts it, contradicts it explicitly. For this contrast is the trace within language of a foundational truth-event.

It can easily be shown that the same applies for 'real' numbers, or for 'imaginary' numbers. Even Cantor's reason for calling the ordinals beginning with ω 'transfinite' numbers becomes less and less obscure for us now, connected as it is to his mindfulness of offending the sanctity of the Infinite with his invention.

The frequency in number-theory of a gap between the trace of a nomination and the sediments of signification indicates that the thinking of number is a true *evental site*: it represents in mathematics a zone of singular precarity and sensitivity, struck regularly by the excess of an event that language and established knowledges consider destitute of signification, and whose destiny can only be sustained by means of a poetic and supernumerary nomination.

And this is because number is, amongst the forms of being, that one which opens onto our thought by way of its organisation (see **10.20**). Which means that everything excessive that thought *encounters* in number, everything that interrupts the regime of its being by way of an evental caesura, has immediate disorganising effects for thought.

12.6. My doctrine of Number, even if my terminology and the echo I give it in philosophical thought are very different things, is nevertheless substantially that of 'surreal numbers' invented by J. H. Conway in the seventies (see **1.7**).[1] I make no claim at all to having produced anything new of a strictly mathematical order. Why, then, change 'surreal number' to just 'Number', with a capital N?

It is basically a poetical disagreement. The nomination proposed by Conway seems to me rather too narrow; let's say that it belongs to an oneiric genre ('surreal' obviously suggesting 'surrealist'), whereas the excessive nature of the discovery in my view demands the majestic genre of the epic, something capable of conveying the unanticipated royal arrival of Number as such.

More technically, it seems to me that 'surreal' remains caught within the notion – all too highly charged with meanings – of a *continuity through successive widenings*. The adjective 'surreal' seems to suggest itself because these new numbers 'contain' the real numbers (as they contain the ordinals); as if the new space conquered was an extension of the old. In his book, Gonshor (see **1.7**), seeking to make propaganda for the surreals, declares that 'we now know the exciting fact that the surreals form a field containing both the reals and the ordinals.'[2] But what is *exciting* in the discovery, at least for the philosopher, goes well beyond this algebraic collection of reals and ordinals. It relates rather to a complete reinterpretation of the very idea of number, to the possibility of finally thinking number as a unified figure of multiple–being. That reals and ordinals arise within this figure is the least of the matter, a simple *consequence*. And all the more so given that, along with reals and ordinals, the misnamed 'surreals' contain an infinitely infinite throng of numbers whose existence no one has conceived of before, and which retroactively make *our* historical numbers seem like a miniscule deduction from all those abundant varieties of numerical being. To give just one example: surreal numbers permit a complete doctrine not only of infinitesimal numbers, but of an infinity of infinitely small numbers, describing a 'downwards' numerical swarming just as vast as that which the ordinals describe 'upwards'.

To use a political image: the nomination 'surreal' seems to me to be marked by that caution, by that attachment to old significations, that characterises a certain 'reformist' reserve when confronted with the event. Now, I think – I wager – that we must adopt the language of rupture here, the 'revolutionary' language. I will say therefore that what takes place here is nothing less than the advent to our thought of Number.

Ultimately the capitalisation of Number does not so much distinguish the genera from the species subsumed to it (whole numbers, rational numbers, real numbers, ordinal numbers, infinitesimal numbers, etc.) – although it does indeed activate such a distinction – as it emphasises the gap between a nomination (here at last is Number) and the diverse significations that, having once been nominations themselves, have become the names of numbers.

12.7. Making thus our wager on the word Number, let us try to legitimise the definition: 'A Number is constituted by the conjoint givenness of an ordinal and a part of that ordinal.'

The ordinals are the ontological schema of the natural multiple. *An* ordinal is a consistent natural unity, counted for one in the ontological situation (set theory). These unities (in the non-numerical sense of the pure and simple consistency of the multiple, of the 'gathering together' of the multiples that constitute it, or belong to its presentation) provide the *material* of Number, that on the basis of which there is Number, or more precisely that within which Number operates a *section*.[3] The simplest way to think about this is to consider that *a* Number extracts a form from its natural ordinal material, as a part, piece or fragment of it, a consistent unit of this material: *an* ordinal.

12.8. Because of their antiquity, their universality, their simplicity (which in fact masks a formidable complexity in the detail), the natural whole numbers will be our guide. We have seen (chapter 11) that, thought according to their being, natural whole numbers are nothing but a particular section in the infinitely infinite domain of ordinals: the section that retains only the initial point of being of this domain (the void) and the 'first' successions, bounded externally by ω, the first limit ordinal. Or that the natural whole numbers extract and isolate, in the boundless fabric of natural multiples, only that which is *finite*.

Why not continue in the same way? It is certainly more rational uniformly to attach the concept of Number to the ordinals in the mode of a section, than to deploy an anarchical selection of disparate procedures (algebraic, topological, set-theoretical . . . , see **1.13**).

Of course, we must be sure this is *possible*. 'Possible' meaning what? That in this way we can find *our* familiar numbers. It would certainly be arbitrary simply to impose, in the name of ontological simplicity, a concept of Number which would not subsume either the rational numbers or the real numbers. But if Number, as a section in a natural multiple, defines whole numbers as well as rational (or fractional) numbers, whole negative numbers as well as real numbers, infinitesimals as well as ordinals, then nothing, in my view, can prevail against both the mathematical *unity* and the philosophical *novelty* of such a concept.

Moreover, the properly ontological simplicity of the idea of 'section' confirms that our wager is good. To say that a Number is constituted on the one hand by an ordinal (which is the signature of the Number's belonging to the natural form of presentation), on the other by a part of that ordinal (which is the section as 'formation' in the natural material) is to define Number by putting to work only the most elementary, 'basic' categories of the ontology of the multiple.

12.9. Number will then appear as the mediation between Nature's infinite prodigality of forms of being and that which we are in a position to traverse and to measure. It is that which, at least in a limited domain of its existence, *accords* our thought the capacity to grasp and measure being qua natural being. Something which every physics confirms.

12.10. There is no doubt that Aristotle's language (Matter and Form) is the most eloquent one for transcribing the idea of Number. In particular, it affords us the advantage of installing ourselves within materialist metaphors. This is no negligible advantage when we know that, since Plato, on account of its apparent mystery, Number has been at the heart of all idealist representations of Nature. Up to, and including, what it has become under the law of Capital, what it is today, as I recounted at the beginning of this book: the unthought basis of the ideology of the countable.

Since the section of Number always operates upon an ordinal, it can be said that, given any Number whatsoever, there always exists an ordinal that is its *matter*. 'Matter' here has a very precise meaning. On the one hand, the ordinal is the 'basis' of Number, that from which its form is sectioned. Thus it proceeds from *one* ordinal, from which an extraction is made, that there should be a Number qua principle of this extraction. On the other hand, we know that all the elements of an ordinal are ordinals (see 8.5). If the numericality of Number, *what* it sections, its form, is a part of an ordinal, then, since all the elements of a part of a set are obviously elements of that set, that which sections a Number must also be *entirely composed of ordinals*. 'Matter' this time means *first matter*. When we speak of the constituents of the numerical section, we are speaking exclusively of ordinals. It is an ordinal that is sectioned, and the elements of the section are also ordinals. With regard to the categories of its matter, Number is natural through and through.

12.11. The Aristotelian metaphor is easily extended: we say that the product of the numerical section, in the ordinal that indicates its natural provenance and furnishes its matter, is the *form* of the Number. Number *itself* is rather the gesture of sectioning, which is why it is represented by the pair of its matter (an ordinal) and its form (a part of that ordinal). But in the form is concentrated that by virtue of which Number *escapes* its natural prescription, or at least *might* escape it. Because the form, being any part of an ordinal whatsoever, brings forth, within a natural unity, a multiple which in general *is not* natural.

The form is, simply, a set of ordinals taken from among the elements of an ordinal. This deduction distinguishes a part of the matter. Now, although every ordinal is a set of ordinals (in fact, the set of ordinals which precedes it, **11.2**), *not every set of ordinals is necessarily an ordinal*. An ordinal has no holes; *all* ordinals that precede it belong to it, from the void 0 right up to itself. This is, moreover, why an ordinal is the name of its own 'length'. If, on the other hand, you take any set whatsoever of ordinals, there is a good chance that a great many ordinals will be missing, that the set will be full of holes. It will therefore not itself be an ordinal. Consequently, the form of a Number is usually not an ordinal; only its matter is. As might be expected in a materialist philosophy, it is matter that is homogenous, non-lacunary, regular, and form that is holey, irregular, non-natural. With the form of a Number we generally transgress the limits of natural being, even if its material is always extracted from within those limits.

12.12. If the form is a part extracted from an ordinal by the section which is Number – a (usually non-natural) subset of a natural set – then it leaves a remainder; there is something like the leftover cuttings from the sculpting of the form in the ordinal–matter. This remainder is made up of those elements of the initial ordinal that are not elements of the form of the Number, the portion of the matter that is not taken up in the form. We call this the *residue* of the Number.

Just like the form, the residue of a Number is a multiple made of ordinals. And, again just like the form, it is usually not an ordinal (it would be somewhat paradoxical if the residue was natural; it *is* so, nevertheless, in the specific case where the form cuts *all* ordinals out of the ordinal–matter without exception, starting from ordinal W). The residue is obtained by the simple *difference* between the Matter and the Form.

It might be objected that, in that case, form and residue are interchangeable. And, in a certain sense, that is the case. Contemporary art has blindly thought this ambiguity in the composition of Number, by exhibiting as new works the residue of works of art whose form is outdated. What will ultimately discriminate between the residue and the form of a given Number, though, will relate to the law of *order* over Numbers, a law we shall study in chapter 13.

Note once more that taking the form and residue together – the union of the form and the residue – restores the matter, that is, the ordinal we began with. The set-theoretical triplet of matter, form and residue is all there is to the numerical section.

12.13. Armed with these remarks, we can outline our programme of investigation into the concept of Number:

- Study what it is that makes the difference between two Numbers, and understand the law of order that serialises them and without which we would not, finite subjects that we are, have any hope of progressing in our knowledge of them.
- Reconstitute algebra, the operational dimension of Number (addition, multiplication, etc.), without which, constrained above all as we are by the ideology of the countable, no one would believe that Number is a number. We will always hold firm to the point that the being of Number precedes operations, that Number is above all a thinking, on the basis of Nature, of a section that extracts a form from a natural unity thinkable as the matter of Number.

- Find again, in the infinitely infinite swarming of Numbers, in the incredible prodigality of being in numerical form, our historical numbers: natural whole numbers, relative whole numbers (negative numbers), rationals (fractional numbers), reals, ordinals . . .
- Show that there exist infinitely more Numbers than we can know or can handle, that our historical numericality is most impoverished compared to the excess of being in Numbers.
- Make sure, in this way, both that Number opens an authentic space for thought and that this thought explains in terms of effective operations only a minute part of all the types of Numbers of which multiple–being – as coupled to thought by set-theoretical ontology – is *capable*.

12.14. This programme accomplished, we will taste the bitter joy of Number, in both its thinkable and its unthinkable aspects. Number will be entrusted to being, and we will be able to turn ourselves toward the numberless effects of the event.

Additional Notes on Sets of Ordinals

N1. The concept of Number makes central use of the concept of 'part of an ordinal'; that is to say, of the concept of an arbitary set of ordinals extracted from a given ordinal. Some remarks must be made concerning the correct treatment of the notion 'set of ordinals', which incorporates that of a 'part of an ordinal', since all the elements of an ordinal are ordinals.

N2. For a set of ordinals to be an ordinal it is necessary and sufficient, as we have noted, that it should have no holes, that no ordinal should disrupt the chain of belonging that binds the ordinals to each other up to the ordinal under consideration.

N3. Since belonging is a total order over the ordinals, *every set of ordinals is totally ordered by belonging*. And this is the case whether or not it has holes. If X is a set of ordinals and x_1 and x_2 are elements of this set, then it is always the case either that $x_1 \in x_2$, or $x_2 \in x_1$, or $x_1 = x_2$. Thus the form and the residue of a Number are totally ordered by belonging, just as its matter is. What makes the form and the residue unnatural are the holes in them, not their order. The universal intrication of natural presentation prescribes its law to all the components of a Number. But what subtracts most Numbers from the strictly natural domain of being resides in the lacks that affect their form (and therefore their residue). A Number is non-natural in so far as its natural fabric is perforated.

N4. Every set of ordinals has a minimal element. This results once more from that important law of natural multiples, the principle of minimality (see **8.10**). Take X, a defined set of ordinals; and P, the property 'belonging to X'. If there exists any ordinal that possesses the property (it is sufficient for this that X should not be empty), then there exists a smallest ordinal that possesses it. It is this smallest ordinal that is the *minimal* element of X: it belongs to all the ordinals of X, but no ordinal of X belongs to it.

The existence of a minimal element has nothing to do with whether or not the set has any holes. Therefore one can always speak of the minimal element of the form of a Number, or of the minimal element of its residue. As to the minimal element of its matter, this is always the empty set 0, since the matter is an ordinal.

N5. We must be very careful on the other hand to observe that *a given set of ordinals does not always have a maximal element*. We have already seen that a limit ordinal (which is a set of ordinals) has no maximal element (**9.14**). A fortiori any set whatsoever of ordinals may very well be infinitely 'open', containing no element that dominates all the others.

N6. However, there always exists an *upper bound* of a set X of ordinals. By 'upper bound' we understand the smallest ordinal to be larger than every ordinal in X. Here again, the existence of an upper bound is guaranteed by the principle of minimality. Let P be the property 'being larger than all the ordinals that belong to X'. There certainly exists an ordinal that possesses this property, unless X is equivalent to the set of all the ordinals, which is inconsistent. Therefore there exists a smallest ordinal which possesses the property P; it is the smallest ordinal to be larger than all the elements of X, and thus it is the upper bound of X. We will denote this upper bound by sup(X).

N7. If a proper part of an ordinal (a part which is not the ordinal itself, a truly partial part) is an ordinal, then it *belongs to* the initial ordinal.

We have known for a long time that the converse principle is part of the definition of ordinals. They are transitive, and so every ordinal that belongs to an ordinal is also a part of it. We now want to show that every *ordinal* which is a proper part of an ordinal belongs to it. This comes back to saying that, between ordinals, the order of belonging *is equivalent to the order of inclusion*.

Suppose an ordinal W_1 is a part of ordinal W_2: $W_1 \subset W_2$. Since belonging is a total order over the ordinals, and since W_1 is different from W_2 (it is a *proper* part of W_2), there are two possibilities:

1 Either $W_1 \in W_2$, and the theorem is true, the ordinal W_1 which is included in W_2 belongs to it, the part is also an element.
2 Or $W_2 \in W_1$. But, since W_1 is transitive, that would mean that $W_2 \subset W_1$. Now we know that $W_1 \subset W_2$. If one set is included in another, and the other included in it, then they are equal, as is intuitively obvious, and as the reader can prove in one line. Now, W_1 cannot be equal to W_2, since it is a proper part of it. Thus the first case must hold, and the theorem is proved.

So it is the same thing, when dealing with ordinals, to say that one belongs to the other, and to say that one is included in the other. In other words, if an ordinal represents (as a part) another ordinal, then it also presents it (as an element). Which does not prevent an ordinal from having some parts *which are not elements*. These parts will simply not be ordinals either. This would be the case, for example, with holey, lacunary sets, sets which begin in the middle of an ordinal chain or only present separated elements, etc. In fact it is generally the case with the form of a Number.

If, however, the form of a Number is an ordinal, then it follows from the preceding arguments that not only is it a part of the matter (the initial ordinal), but also an element of it. Then the form is of a peculiar kind, like a 'corpuscle' of matter. In such cases, Number is less a representation extracted from Nature than a simple natural presentation.

13

Difference and Order
of Numbers

13.1. A Number is entirely determined by its matter (the ordinal from which its form is extracted) and its form. The residue is obtained by taking the difference between the matter and the form. Because of this, it is often convenient, as we have said, to write a Number in the form $N = (W, F(N))$, where W is the ordinal–matter and $F(N)$ the form. The residue $R(N)$, is equal to $W - F(N)$.

Given these conditions, how can we think the *difference* between two Numbers? It is natural to posit that they are identical if they have the same matter and the same form. If they are not identical, this could be:

- because they do not have the same matter. Take W_1 the ordinal–matter of one, and W_2 the ordinal–matter of the other. Two ordinals like W_1 and W_2 are ordered by belonging, but also, as we have seen (N7), by inclusion: either $W_1 \subset W_2$, or $W_2 \subset W_1$. Thus we can say that in this case what differentiates W_1 from W_2 is the set $W_2 - W_1$, or $W_1 - W_2$. Since all elements of an ordinal are ordinals, we can also say that what differentiates W_1 from W_2 – and thus the numbers N_1 and N_2, of which these ordinals are the matters – are the ordinals which are elements of W_2 but not of W_1 (if $W_1 \subset W_2$) or elements of W_1 but not of W_2 (if $W_2 \subset W_1$);

- because, having the same matter, they do not have the same form. In this case, there are elements (and therefore ordinals, since the first matter of a Number is composed of ordinals in its three components, matter, form and residue) that are in the form of one but not in the form of the other. But, since the matter is the same, every element of the form of one which is not in the form of the other is in its residue: if $W \in F(N_1)$ and $W \notin F(N_2)$, then $W \in R(N_2)$. What differentiates the two Numbers N_1 and N_2 is the set of ordinals that are in the form of one and in the residue of the other.

We can see then that the difference between two Numbers can be understood in terms of ordinals. If an ordinal is in the matter of one and not in that of the other, or if it is in the form of one and in the residue of the other, it *makes a difference* between the two Numbers.

13.2. Take any two Numbers whatsoever. We will say that an ordinal *w discriminates between* these two Numbers if it is in the matter of one and not in that of the other, or if it is in the form of one and in the residue of the other (which implies that it is in the matter of both, since form and residue are both parts of matter).

13.3. Let's take an example: Let N_1 be the Number $(2,(0))$ whose ordinal–matter is 2 and whose form is (0). It is certainly a Number, since 2 is an ordinal (it is the finite ordinal whose being is $(0,(0))$, see **11.5**) and the singleton of 0, denoted by (0), is a part of that ordinal **(7.11)**. This Number N_1 has, for matter, the ordinal 2, and, for form, the part (0).

Now let N_2 be the Number $(\omega, 2)$. Once again it is a Number, since ω is an ordinal (the first limit ordinal) and the ordinal 2, which is an element of ω, is also a part of it (transitivity of ordinals). This Number N_2 has for its matter ω and for its form the part $(0,(0)) = 2$.

The ordinal ω *does not discriminate between* the Numbers N_1 and N_2. Indeed, ω is certainly not in the matter of N_1 (which is 2, a finite successor ordinal), but neither is it in the matter of N_2, because this matter is ω, and we know that no set belongs to itself: it cannot be that $\omega \in \omega$.

The ordinal 0 (the empty set) *does not discriminate between* the Numbers N_1 and N_2 *either*. In fact, it is in the form of both. The form of N_1 is the singleton (0), of which 0 is the only element. So 0 is an element of this form. And, on the other hand, 0 is an element of the ordinal 2, which is the form of N_2. Thus 0 is also in the form of N_2.

However, the ordinal (0) (which is the whole number 1) *does discriminate between* the Numbers N_1 and N_2: (0) is an element of the ordinal 2, and thus belongs to the form of N_2. But it cannot belong to the form of N_1, which is precisely (0), since the self-belonging $(0) \in (0)$ is impossible. Given that (0) is an element of the matter of N_1 (which is the ordinal 2), since it is not in its form, it must be in its residue.

13.4. Given two Numbers and any ordinal whatsoever, it is always possible to say whether this ordinal discriminates between the two Numbers or not. If N_1 and N_2 are Numbers, the property 'discriminating between N_1 and N_2' is well-defined.

But if there is an ordinal that discriminates between N_1 and N_2 (that is, if N_1 and N_2 are different), then in virtue of the principle of minimality – which we have constantly made use of because it is a fundamental law of natural multiples (see **8.10**) – there is one unique smallest ordinal which discriminates between them. Or, if you like, a minimal ordinal for the property 'discriminating between the Numbers N_1 and N_2'.

13.5. An *extremely important definition*: **The smallest ordinal to discriminate between two Numbers is called their *discriminant*.**

The interesting thing about the concept of discriminant is the following: it brings the idea of the difference between two Numbers down to a matter of *one single ordinal*. This 'minimal point' of differentiation allows a *local* rather than global treatment of the comparison between two Numbers. The existence of a discriminant suffices for us to conclude that two Numbers are different.

13.6. One more example. Take the two Numbers N_1 and N_2 from the example above (**13.3**), $N_1 = (2,(0))$ and $N_2 = (\omega,2)$. What is their discriminant?

- We have seen that·0 does not discriminate between N_1 and N_2.
- We have also seen that (0) discriminates between them. Since the only ordinal smaller than (0) is 0, which does not discriminate between N_1 and N_2, (0) is definitely the smallest ordinal that discriminates between them. We therefore say that (0) is the discriminant of the Numbers $(2,(0))$ and $(\omega,2)$.

Note the location of the discriminant: it is in the matter of the two Numbers N_1 and N_2, but is in the form of N_2 and in the residue of N_1.

Now consider the following two Numbers (S(W) denotes the successor of the ordinal W, see **9.5**):

- $N_3 = (S(\omega),\omega)$. Its matter is the ordinal $S(\omega)$, its form ω itself. The latter is a part of $S(\omega)$, since every ordinal is an element of its successor, and every element of an ordinal is a part of it (transitivity).
- $N_4 = (S(S(\omega)),\omega)$. Its matter is the successor of the successor of ω, its form is ω.

What is the discriminant of N_3 and N_4? These two Numbers have the same form, that is ω, but extracted from different matters, $S(\omega)$ and $S(S(\omega))$. In summary, in these numbers everything is exactly the same *up to the ordinal S(ω)*. This ordinal is in the matter of N_4, since $S(\omega) \in S(S(\omega))$, but it is not in the matter of N_3, since $S(\omega) \notin S(\omega)$. Thus the ordinal $S(\omega)$ is the smallest ordinal to make a difference between N_3 and N_4; it is the discriminant of these two Numbers.

Note once again the location of the discriminant of N_3 and N_4: $S(\omega)$ is not in the matter of N_3, but is in that of N_4. Meanwhile, this time *it is not in the form of N₄*, which is ω. It is therefore in its residue.

The combination between the ordinal punctuality of the discriminant and its location in the Numbers compared will give us the key to the concept of order in the boundless domain of Numbers.

13.7. Let's give all of this a slightly stricter form.

The *location of an ordinal w with regard to Number N*, written $L(w,N)$, is the position that it occupies with regard to the three dimensions of the numerical section carried out by the Number N: matter, form, residue. There are obviously three locations:

1   Either the ordinal w is not an element of the ordinal W which is the matter of the Number N. In this case we say that it is located 'outside the matter' and we posit that: $L(w,N) = oM(N)$.
2   Or the ordinal w is in the matter W and belongs to the form of the Number. We then posit $L(w,N) = F(N)$.
3   Or the ordinal w is in the matter W, but belongs to the residue of the Number: We then posit $L(w,N) = R(N)$.

When there is no ambiguity as to the number N in question, we might simply use the notation $L(w) = R$, signifying that the location of w (for the number in question, of course) is its belonging to the residue (of that number).

Given a number N, every ordinal can be located for N so long as we allow the location 'outside the matter'.

When an ordinal discriminates between two numbers N_1 and N_2 (see **13.2**), it is very simply because its location in the two Numbers is not the same. The table of possible locations for an ordinal *w which discriminates between the two Numbers* is as follows (using oM, F and R to denote the locations):

L (w,N_1)	L (w,N_2)
F	R
F	oM
R	F
R	oM
oM	F
oM	R

The discriminant of N_1 and N_2, being the smallest ordinal to discriminate between them, necessarily responds to one of the 'pairs' of locations indicated in the table. For example, if it is in the residue of N_1, it must be in the form of N_2 or outside the matter of N_2, etc.

13.8. Definition of order over Numbers
Take two Numbers N_1 and N_2 and their discriminant w (if necessary, reread **13.4–13.6**, given that the concept of discriminant is central).

We say that N_1 is smaller than N_2, written $N_1 < N_2$, if the location of the discriminant w for the Numbers N_1 and N_2 satisfies one of the three following cases:

1 Either $L(w,N_1) = R(N_1)$, and $L(w,N_2) = F(N_2)$: *the discriminant is in the residue of N_1 and in the form of N_2.*
2 Or $L(w,N_1) = oM(N_1)$, and $L(w,N_2) = F(N_2)$: *the discriminant is outside the matter of N_1 and in the form of N_2.*
3 Or L $(w,N_1) = R(N_1)$ and $L(w,N_2) = oM(N_2)$: *the discriminant is in the residue of N_1 and outside the matter of N_2.*

Compare these three cases carefully with what the table in **13.7** indicates as to the possible locations of the discriminant of two Numbers.

13.9. It is not immediately evident that the relation $N_1 < N_2$ is one of order. But, even before establishing that this is the case, we can reveal the characteristics of this relation.

The discriminant gathers into one point (one ordinal) the concept of difference between two Numbers. The order introduced here depends on the location of this point, and therefore on a sort of *topology of difference*. Since, in the gesture of sectioning that constitutes every Number N, the 'positive' numericality – that which this gesture extracts from matter – is the form, we will always consider that, *if* the discriminant of two Numbers is in the form of one, this number is 'larger' than the other. In the other, of course, the discriminant will either be in the residue or outside the matter.

Conversely, the residue of a Number is the purely passive result of the section of its form, the unintentional remainder of the numeric gesture. It is that which Number as gesture *leaves to matter*. If the discriminant of two Numbers is in the residue of one of them, we will always consider this Number to be 'smaller' than the other. In the other, the discriminant will be in the form, or outside the matter.

13.10. An apparently paradoxical consequence of this conception, which determines *all order* on the basis of the active superiority of form – thought as the numericality of Number – over residue – thought as passive inverse – is that a number N_1 is said to be smaller than N_2 if their discriminant is in the residue of N_1 *and outside the matter of N_2*. The 'paradox' results from the assumption that the position 'outside matter' is completely unaffected by the numerical gesture, being neither in its form nor in its residue. Isn't it even more passive then, even less involved in the numerical extraction of the form, than an ordinal which is in the residue, and which therefore at least figures in the matter of the Number? Isn't the location oM a figure of *nothing* in relation to Number, an ontologically 'inferior' figure to the passive figure of the residue?

13.11. This sense of 'paradox' misses an essential point, which is that *the 'outside matter' position includes the matter itself*, since an ordinal W is not an element of itself. There is no reason to suppose that the matter is 'indifferent' to the gesture of Number: it is its primary 'given', that on the basis of which there is Number; the natural multiple whose being is exposed to the numerical section. And it is *always* the matter itself of one of the two Numbers that is at stake when the discriminant is located 'oM' for one of them.

If the discriminant of N_1 and N_2 is, say, outside the matter of N_2, then it is in the matter of N_1 (in its form, or in its residue). If not – if it were outside the matter of both Numbers – it could not discriminate between them. Therefore the discriminant must indeed be the smallest ordinal in the matter of N_1 and outside the matter of N_2. Evidently this means *that the ordinal W_1 which is the matter of N_1 is larger than the ordinal W_2 which is the matter of N_2.* If not, there could be no ordinal in W_1 that was not in W_2, since the elements of an ordinal are all the ordinals that precede it (see **11.2**). This means that $W_2 \in W_1$ (the order-relation over the ordinals is belonging). But W_2 itself is the smallest ordinal that does not belong to W_2, since all the ordinals smaller than W_2 are, precisely, the elements of W_2. And so, ultimately, W_2 is in W_1 and is the smallest ordinal not to be in W_2. It is the smallest ordinal to be in W_1 (the matter of N_1), and not in W_2 (the matter of N_2), and therefore outside the matter of N_2. W_2, *the matter of N_2, is the discriminant of N_1 and N_2.*

This demonstration has a general validity: whenever we say that N_1 is 'smaller' than N_2, or that $N_1 < N_2$, because the discriminant of N_1 and N_2 is in the residue of N_1 and outside the matter of N_2, *this also means that the discriminant in question is the ordinal–matter of N_2.* And this relation is legitimate because the matter of a Number, the one-ordinal in which the numerical section operates, is a primordial donation of being ontologically superior to the passivity of the residue.

It is therefore philosophically well-founded to put the locations in the order $R < oM < F$: the form, affirmative numericality of the section, is superior to what is outside matter, which is itself superior to the passivity of the residue because in reality this 'outside of matter' is the matter itself, integrally counted for one as an ordinal.

The relation $N_1 < N_2$ founded on our three cases (the discriminant in $R(N_1)$ and $F(N_2)$; the discriminant in $R(N_1)$ and $oM(N_2)$; the discriminant in $oM(N_1)$ and in $F(N_2)$) describes a hierarchy, founded in the being of Number as the sectioning of a form in natural matter.

13.12. What remains now is to establish that the relation $N_1 < N_2$ is truly an order-relation, in the mathematical sense of the term: that it *serialises* Numbers. This amounts to responding positively to three questions:

1 Is the relation *total?*[1] Or: given any two different Numbers N_1 and N_2, is it always the case that $N_1 < N_2$ or $N_2 < N_1$?
2 Is the relation *non-reflexive?* Or: is it impossible that $N_1 < N_1$?
3 Is the relation *transitive?* Or: from the relations $N_1 < N_2$ and $N_2 < N_3$ does it necessarily follow that $N_1 < N_3$?

If we prove these three points, we will have brought the philosophical legitimacy established in **13.11** to coincide with mathematical (ontological) legitimacy. Or, rather, with regard to the order of Numbers, we will have obtained the situation in which we have been continuously striving to remain: where what is said under the sign of the philosophical statement 'mathematics is ontology' remains in harmony with what is said under the sign of mathematical inferences themselves. Or where the interpretation of mathematics as science of being qua being draws its contact with the real from the effective thoughts of such a science.

13.13. The relation < is *total*

Look one last time at the table of cases of inequality for $N_1 < N_2$. This table fixes the location of the discriminant of N_1 and N_2.

	N_1	N_2
Case 1	R	F
Case 2	oM	F
Case 3	R	oM

To demonstrate that the relation is total is to show that, given two different Numbers, one of them is always 'smaller' than the other.

Take two randomly selected numbers N_3 and N_4, and w, the ordinal which is their discriminant. Three cases are possible:

1   The discriminant w is in the residue of N_3. Then:
   (a)   either it is in the form of N_4, and (see the table) $N_3 < N_4$;
   (b)   or it is outside the matter of N_4, and (*idem*) $N_3 < N_4$.
2   The discriminant w is in the form of N_3. Then:
   (a)   either it is in the residue of N_4, and (*idem*) $N_4 < N_3$;
   (b)   or it is outside the matter of N_4, and (*idem*) $N_4 < N_3$.
3   The discriminant is outside the matter of N_3. Then:
   (a)   either it is in the form of N_4, and (*idem*) $N_3 < N_4$;
   (b)   or it is in the residue of N_4, and (*idem*) $N_4 < N_3$.

Having exhaustively enumerated all the possible cases, we see that the relation < between Numbers N_3 and N_4 is always defined. The relation really is total in the domain of Numbers, there are no two different Numbers *not related* by <.

13.14. It is good to get into the habit of thinking through the inequalities between Numbers more rapidly. For example we could say: if the discriminant w is in the residue of one of the two Numbers, the table (cases 1 and 3, which are the only possibilities) shows that this Number is smaller than the other. If w is in the form of one of the two Numbers, the table (cases 1 and 2, the only possibilities) shows that this number is larger than the other. Apparently we have left to one side the case where w is outside the matter of one of the Numbers. Not so, because then it would necessarily be in the residue or the form of the other (if it was outside the matter of both, it would not discriminate between them), and we are referred back to one of the preceding cases.

To compare two numbers according to the relation <, we therefore proceed as follows: firstly we check whether the discriminant is in the form of one of them: if so, we conclude immediately that this Number is the largest. If not, we check whether it is in the residue of one of the two: if so, we conclude that that Number is the smallest. The work is done, no other case is possible.

13.15. The relation < is *irreflexive*
This point is trivial. It cannot be that $N_1 < N_1$, since the relation is founded on the existence and location of a discriminant, which cannot exist 'between' N_1 and itself.

13.16.
To exercise ourselves in the comparison of Numbers using the < relation, let's take up the examples from **13.6** once more. We had, adopting the notation by the pair of matter and form, the four following Numbers:

$$N_1 = (2,(0))$$
$$N_2 = (\omega,2)$$
$$N_3 = (S(\omega),\omega)$$
$$N_4 = (S(S(\omega)),\omega)$$

The discriminant of N_1 and N_2 is (0). It is in the residue of N_1, and in the form of N_2. So $N_1 < N_2$.

The discriminant of N_3 and N_4 is $S(\omega)$. It is outside the matter of N_3, and in the residue of N_4. So $N_4 < N_3$.

The discriminant of N_3 and N_1 is (0), which is in the residue of N_1 and in the form of N_3. So $N_1 < N_3$.

The discriminant of N_4 and N_2 is 2 (why?). It is in the residue of N_2 and in the form of N_4. So $N_2 < N_4$.

The reader can study the remaining comparisons on their own account.

It will be remarked that it is not simply because the *matter* of a Number is 'larger' that that Number is larger. Thus N_4 has for ordinal–matter the successor of the successor of ω, which is larger than the successor of ω, the matter of N_3. Nevertheless, $N_4 < N_3$.

What is still more remarkable in this example is that the form of N_3 and of N_4 is the same (it is ω). Thus we have the following 'law': *if the form stays the same whilst the matter grows, the Number gets smaller*. It is quite straightforward to demonstrate the general case. Take a Number N_1 (W_1,X) and a Number N_2 (W_2,X), where $W_1 \in W_2$; and X the same set of ordinals (which is a common part of W_1 and W_2). The discriminant of these two Numbers cannot be found in the form of either of them, since they have the same form, X: an element of X cannot discriminate between them (it even has a location for N_1 and N_2, namely the form). It is therefore in the residue of one, and outside the matter of the other. It is clear that this discriminant is none other than W_1, which is the smallest ordinal not to belong to W_1, and which is in W_2, since $W_1 \in W_2$. Now W_1 is necessarily in the residue of N_2 (since it belongs to its matter, but not to X, its form), and outside the matter of N_1. Therefore it is indeed the case that $N_2 < N_1$.

This process suggests a comparison between the Number (W,X) and the relation $\frac{X}{W}$. We know that such a relation *diminishes* when its denominator W grows. But be warned: this is only a distant analogy, because $\frac{X}{W}$ means nothing here.

All the same we can show, inspired by this analogy, that, *if the matter remains the same whilst the form gets larger – so that the old form X is included in the new form X′ – then the Number gets larger*. This time, it is the enlargement of the 'numerator' that enlarges the 'relation'. I leave the details of the demonstration of this to the reader. Suffice to say that the discriminant is the smallest ordinal to belong to X′ and not to X; so it is in the form of the second Number and in the residue of the first, and therefore the second is larger.

These observations are philosophically well-founded. What does it mean, in fact, to produce the same form from a larger matter? That the gesture of the numerical section did not manage to extract from a vast matter (that of a larger ordinal) any more of a form than could have been obtained with a smaller matter. The gesture was thus less concentrated, less elegant, less effective. It is quite legitimate that the Number which marks this gesture should be held for inferior. The converse also follows: to obtain a more widely deployed form, containing all the elements of the first and more, with the same initial matter, requires a more efficient gesture of sectioning. It is quite right that this should be marked by a superior Number.

The relation < does indeed express in the mathematical field the ontologically rational dispositions of the comparison of Numbers.

13.17. The relation < is *transitive*

This is a question of proving that, given three Numbers, N_1, N_2 and N_3, if $N_1 < N_2$ and $N_2 < N_3$, then $N_1 < N_3$. Obviously everything hinges on the location of the discriminants. We shall write the discriminant of N_1 and N_2 as $w(1,2)$, that of N_2 and N_3 as $w(2,3)$, and that of N_1 and N_3 as $w(1,3)$.

(a)   First step. An ordinal smaller than $w(1,2)$ *and* $w(2,3)$ does not discriminate between N_1 and N_3.

The discriminant of two Numbers is *the smallest* ordinal that discriminates between those two Numbers (in the order of ordinals, which is belonging).

If an ordinal W is smaller than $w(1,2)$, it doesn't discriminate between N_1 and N_2. Its location (F,R, or oM) is the same in N_1 and in N_2. Equally, if it is smaller than $w(2,3)$, it doesn't discriminate between N_2 or N_3 either – its location is the same in N_2 and N_3. Ultimately, therefore, its location must be the same in N_1, in N_2 and in N_3, and it does not discriminate between N_1 and N_3.

(b)   Conclusion of the first step: $w(1,3)$, which obviously discriminates between N_1 and N_3, cannot be smaller than $w(1,2)$ *and $w(2,3)$. It is therefore at least equal to the smaller of the two.*

(c)   *Second step.* The smallest of the two ordinals $w(1,2)$ and $w(2,3)$ discriminates between N_1 and N_3.

For convenience of exposition, we will suppose that the smallest is $w(1,2)$ (the reasoning would be exactly the same if it was $w(2,3)$; confirming this would be an excellent exercise for the reader). Since $w(1,2)$ discriminates between N_1 and N_2, its location in N_1 is different from its location in N_2. But, since it is smaller than $w(2,3)$, it does *not* discriminate between N_2 and N_3, since $w(2,3)$ – discriminant of N_2 and N_3 – is the smallest ordinal that discriminates between these two Numbers. Therefore the location of $w(1,2)$ in N_2 and N_3 is *the same*. Since its location in N_2 differs from that in N_1, if it is the same in N_3 as in N_2, its location in N_3 also differs from its location in N_1. So $w(1,2)$ discriminates between N_1 and N_3.

(d)   *Third step. $w(1,3)$, the discriminant of N_1 and N_3, is actually equal to the smallest of the ordinals $w(1,2)$ and $w(2,3)$.*

We have seen that $w(1,3)$ must be at least equal to the smallest of the two ordinals $w(1,2)$ and $w(2,3)$ (first step). We supposed $w(1,2)$ to be the smallest. Thus $w(1,3)$ is at least equal to $w(1,2)$. Now $w(1,2)$ discriminates between N_1 and N_3 (second step). Since $w(1,3)$ is the discriminant of N_1 and N_3, and thus the smallest ordinal to discriminate between them, and since it cannot be smaller than $w(1,2)$, which discriminates between N_1 and N_2, *it is equal to $w(1,2)$. So $w(1,3) = w(1,2)$*.

(e)   An aside: if we were to suppose the opposite hypothesis, that $w(2,3)$ is smaller, we would find that $w(1,3) = w(2,3)$, for the same reasons.

(f)   *Fourth step*, the conclusive step. We have discovered that $w(1,3) = w(1,2)$. This can be expressed as follows: $w(1,2)$, discriminant of N_1 and N_2, is also the discriminant of N_1 and N_3.

Now, we know that $N_1 < N_2$. So we know there are two possible locations for the discriminant $w(1,2)$ in N_1, the smaller of the two Numbers:

1   Either $w(1,2)$ is in the residue of N_1. But then, since it is also the discriminant of N_1 and N_3, its position in the residue of N_1 leads us to conclude that $N_1 < N_3$ (on this point, see **13.14**).
2   Or $w(1,2)$ is outside the matter of N_1. It must then be in the form of N_2, for the usual reason that $N_1 < N_2$. But $w(1,2)$, which is smaller than $w(2,3)$, does not discriminate between N_2 and N_3. Therefore it is also in the form of N_3. And, since it is the discriminant of N_1 and N_3, being outside the matter of N_1 and in the form of N_3, once again $N_1 < N_3$.

So we have proved that, if $N_1 < N_2$ and $N_2 < N_3$, then $N_1 < N_3$. We have even discovered, as a bonus, a still finer result: the discriminant of N_1 and N_3 is equal to the smallest of the discriminants of N_1 and N_2, and of N_2 and N_3.

13.18. Dialogue with a tenacious reader, on the subject of the preceding demonstration:

THE READER: You suppose from start to finish that the discriminant of N_1 and N_3 *exists*. It's not so obvious. I can well see that the discriminant of N_1 and N_2 exists, since we know that $N_1 < N_2$. The same for that of N_2 and N_3. But it could well be that in the end $N_1 = N_3$, and in that case there would be no discriminant $w(1,3)$. The relation would be circular: $N_1 < N_2 < N_1$.

ME: But that's absurd: If the discriminant of N_1 and N_2 is located in N_1 and N_2 in such a way that $N_1 < N_2$, it cannot be the case that

$N_2 < N_1$. Therefore N_1 is necessarily different from N_3, and their discriminant exists.

READER: Okay, you've got me. But I'm still not satisfied. In your second step, you suppose that one of the two discriminants $w(1,2)$ and $w(2,3)$ is the smaller of the two. But surely it could quite easily be the case that there is no smallest of the two; for this to be the case it suffices that they be *equal*. And, so that you don't try to pull the wool over my eyes, I'll give an example. Take these three numbers:

1   N_1 is the number $(2,(1))$, which has for its matter 2 and for its form the singleton of 1. I know (I've read you saying so just now) that the ordinal 1 is an element of the ordinal 2 (see **11.5**), and that the singleton of an element is a part (**7.10**). Here we have the pair of an ordinal and of a part of that ordinal, so it's a Number (**12.1**).
2   N_2 is the Number $(0,0)$, which has for matter the empty set, and for form the empty set. It's still a Number though! Because 0 is an ordinal, which serves for the matter, and 0 is a universal part of every set (see **7.9**), including 0 itself, which is, as I know, a set. So 0 is fine as the form too.
3   N_3 is the Number $(2,1)$. You can't refuse me this, because 2 is an ordinal, and 1, being an element of the ordinal 2 (following your **11.5**, as always), is also a part of it, since every ordinal is transitive.

Now, let's see, what do I have? The discriminant of N_1 and N_2 is 0: it's in the residue of N_1, since 0 is an element of the matter 2, but does not figure in the form, the singleton (1), whose only element is 1. And it's outside the matter of N_2, since this matter is 0, of which 0 cannot be an element. Therefore $N_1 < N_2$.

The discriminant of N_2 and N_3 is *also* 0, which is outside the matter of N_2, as we can see, and which is in the form of N_3, since $0 \in 1$. From this we conclude that $N_2 < N_3$.

Here is a concrete test case where $N_1 < N_2$, where $N_2 < N_3$, and where, nevertheless, $w(1,2)$ and $w(2,3)$, to use your notation from the beginning of **13.17**, are equal. Therefore neither is smaller than the other, and your chain of inference is broken.

ME: Very shrewd! You will have to allow me all the same that in the end, in your example, transitivity is confirmed. Because the discriminant of N_1 and N_3 is still 0, which is in the residue of N_1 and in the form of N_3. So it is still the case that $N_1 < N_3$.

READER: I make an objection on a point of principle, and you respond with an empirical remark! My example ruins your general argument, which rests on the fact that one can *always* discern the smallest of the discriminants of N_1 and N_2 and of N_2 and N_3. I have shown you a case where this cannot be done. The fact that transitivity still works for my example might just be chance, since it now seems you have yet to *prove* anything.

ME: You allow my first step, all the same: that $w(1,3)$ cannot be smaller than $w(1,2)$ *and* $w(2,3)$?

READER: With the caveat that the 'and' seems somewhat suspect to me, since it might relate two equal discriminants. See my example: you would be saying that '$w(1,3)$ cannot be smaller than 0 *and* 0', which is ludicrous.

ME: Unless it *was* smaller than 0 . . . But anyhow – if, as in your example, $w(1,2)$ is equal to $w(2,3)$, do you admit that $w(1,3)$ cannot be smaller than $w(1,2)$ *alias* $w(2,3)$? Because no ordinal smaller than this common discriminant can discriminate between N_1, N_2 and N_3.

READER: Obviously.

ME: But $w(1,2)$ discriminates between N_1 and N_2 – its location in N_1 isn't the same as in N_2?

READER: No; how could it be?

ME: And it also discriminates (going by the name of $w(2,3)$, to which it is equal) between N_2 and N_3 – its location is not the same in these two?

READER: That's exactly what I said.

ME: Let's look at these locations more closely. Since $N_1 < N_2$, $w(1,2)$ must either be in the residue or outside the matter of N_1. But can it be outside the matter?

READER: (after some time thinking) No. Because, if it were outside the matter of N_1, it would have to be in the form of N_2, since $N_1 < N_2$. But, as it is also the discriminant of N_2 and N_3, and $N_2 < N_3$, it cannot be in the form of N_2, as explained in **13.14**. So it is definitely in the residue of N_1 and . . .

ME: . . . outside the matter of N_2, because not in its form. But where is it located in N_3?

READER: (after some time thinking) In the form. Because this $w(1,2)$, which is also $w(2,3)$, is the discriminant of N_2 and N_3. Being outside the matter of N_2, since $N_2 < N_3$, it is in the form of N_3.

ME: Perfect! Here is a $w(1,2)$ which $w(1,3)$ cannot be less than, and which is found in the residue of N_1 and in the form of N_3. Therefore it discriminates between N_1 and N_3. That is to say . . .

READER: Okay, I get it. Already identical to $w(2,3)$, it must also be identical to $w(1,3)$. And this identity means that $N_1 < N_3$, since this common discriminant (of N_1 and N_2, of N_2 and N_3, and of N_1 and N_3) is in the residue of N_1 and in the form of N_3. That works.

ME: It's just as your example says: 0 was the common discriminant of the couples N_1–N_2 and N_2–N_3. It is also the discriminant of N_1 and N_3. And it is located in the residue of N_1, outside the matter of N_2, and in the form of N_3. Which gives us the sequence: $N_1 < N_2 < N_3$.

READER: You must admit that you've had to add quite a bit to your original account.

ME: It is a subsection of the argument, the principle is the same. But in mathematics one cannot skip over anything, for the reason that one *never knows* what one is skipping over.

13.19. Since the relation $<$ is total, irreflexive and transitive, it really is an order-relation in the mathematical sense. We have entirely justified our saying 'N_1 is smaller than N_2' when it is confirmed, by means of the location of the discriminant of N_1 and N_2, that the relation $N_1 < N_2$ is valid.

Thus the universe of Numbers – even if it is, as we shall see, borderless, saturated to an inexpressible degree, of a density with regard to which the celebrated 'continuum' is thin and lacunary – can nevertheless be comprehended wholly under the serial law of an order.

The additional fact that this order can be designated solely by the examination of the location of *an* ordinal (the discriminant) with regard to three possible sites (F, R and oM) indicates a simplicity that is reassuring as to *our* capacity to think the universe of Numbers.

It is striking that, given its combination of a logic of minimality (the discriminant, the smallest ordinal to mark the difference of two Numbers) and a logic of positions (the three components of the numerical section), this order appears to be allied with lexicographical order. In fact, it is presented as such in purely mathematical expositions.[2]

Now, lexicographical order, which organises words by recourse to an alphabet of the phonic or scriptural unities that compose them, touches on the distinction, so important in Lacan, between the signifier and the letter.[3] In reality, Number is indeed like a signifier, whose internal 'positions' are the three locations – matter, form and residue – and whose letters are the ordinals. This alone permits us to organise something as anarchic as sets of any ordinals whatsoever, ordinal 'words'.

If Number is the medium in which Nature, grasped in its being, opens itself to our thought, this is, without doubt – as the order of Numbers testifies – because, in the section it carries out, we find, under the simple form of one and three, that dialectic of the position and of the letter which has been recognised, since Galileo, as the true terrain of materialism. Nature consents to its profusion within the fiction of a writing system; and we must recognise in Number the most *inscribed* instance of being:

> two fingers
> snap in the abyss, in
> scribblebooks
> a world rushes up, this depends
> on you.[4]

14

The Concept of Sub-Number

14.1. The concept of substructure, and even (in category theory)[1] that of the sub-object, is fundamental for all areas of contemporary mathematics. We know the extreme importance of the determination of subgroups of a group, subspaces of a topological space, etc. A good many of the most profound mathematical theorems of recent years are theorems of decomposition or of presentation: proving that a structure can be presented as a composition of (possibly simpler) substructures, or that a structure is decomposable into a sequence of, or as a product of, pre-defined substructures. The elegance of thought reaches its highest point when one manages completely to 'resolve' a presented axiomatic structure into substructures that are of the same type, but simpler. Finite group theory offers some spectacularly accomplished examples of such resolution.

The underlying idea is as follows: since the 'material' of mathematics is the pure or undifferentiated multiple, structures are inevitably homogenous with structured sets. Mathematical ontology is unitary: there aren't, on the one hand, pre-given 'objects', on the other, structural relations into which these objects enter. Everything can potentially be reduced to a multiple without quality, made of the void alone. Given this fact, it is inevitable that the exercise of thought should consist in reducing complex multiplicities to simpler multiplicities, through the medium of the axiomatic definition of simple and complex. The concept of structure organises this medium: it distinguishes elementary configurations from more intricate configurations. Ultimately the strategic stakes of the thinking of being qua being are to discern – given that every multiple is a multiple *of mul-tiples* (One having no being) – which multiples a presented multiple assures, in its turn, the presentation of. Whence theorems of decom-position, resolution, or presentation.

What a mathematician calls an 'object' is nothing but a multiplicity within which sub-multiplicities are intricated, often in a very opaque fashion. The object is a packet of multiples, whose intrication is an obstacle to thought, and within which must be *separated*, as far as possible, the multiples–regions whose presentational combination is assured by the 'object'. The 'objective' illusion, what we might call the phantasm of the object, relates to the initial distance between the entanglement of multiples and *our* separative access to this entangle-ment through the medium of language. Concepts, axiomatically introduced, determine types of structures, which are the operators of separation and allow us to exhibit such an 'object' as an articulation of substructures, indicating the latency of sub-multiplicities in their relation to the medium of language.

That a structure can be resolved into substructures according to various operators of combination (embedded sequences of subgroups, finite or infinite products of compact spaces, etc.) is the definitive mark, in the *inscribed* strategy of thought, of the fact that what it confronts is being qua being, in the figure of an infinite entanglement of pure multiplicities. A mathematician will say that he has 'thought the object' (or 'understood the problem') when he has mapped the linked immanence of the substructures whose presentative bond is detained, initially in an opaque fashion, by the 'object'. So it is also a question of the decomposition of the object, a putting to death of the phantasm of the object, which is an object only in so far as it resists, through its constitutive entanglement, its resolution into the specific diversity of structures. Thinking by means of substructures *deposes the object* and returns toward being.

14.2. In its commonly accepted usage, *the concept of number is not a concept of the structural type*. One doesn't speak of 'numerical structure' as one speaks of the structure of groups or of vector space. What is called 'number-theory' today is an inconsistent set whose centre of gravity is in fact a certain area of algebra: ring theory and ideals theory. In particular, no concept of sub-number exists, since 'number' doesn't designate a type of structure.

Consequently, since the Greeks, the concept of number has been the principal redoubt of a realist, even empiricist, vision of mathemat-ics. Either number is taken for a 'given' entity, or taken as proof that mathematical nominations have a strictly symbolic or operational value. There is a closing-in-on-itself of the entity 'number', which is linked to its purely algebraic manipulation. Certainly, numbers are combined according to algebraic rules. But it does not at all follow from $7 + 5 = 12$ (whether this statement be analytic or synthetic) that 7 and 5 are 'substructures' of 12. The most tenacious illusion of objectivism resides in the conviction that 7, 5 and 12 are non-decomposable marks, whose serial engenderment assures their consistency.

It would therefore be a great victory for an ontological vision of mathematics to establish the structural character of number, to unbind it from its empirical punctuality, to extract it from the simple form of the object. This programme, which would make of the predicate 'number' a reputable type of pure multiple, would find its most significant moment in the determination of the concept of sub-number. This concept would align numericality with the great structural categories of mathematical thought (group, field, space . . .); categories by means of which thought separates and unbinds the intrications of the pure multiple.

14.3. The set-theoretical presentation of the concept of Number, such as we have worked it through above, authorises a strict definition of the sub-Numbers of a given Number. Better still: as we shall show step by step, *a Number is defined in a univocal manner by its sub-Numbers*. There exists a presentation of Number on the basis of the Numbers that are immanent to it. Thus Number in its turn admits of theorems of decomposition or of presentation. It is *structuralised*.

14.4. The concept of sub-Number

The general idea of the sub-Number is very simple: we obtain a sub-Number of a given Number if we 'partition'[2] this Number at a point of its matter and keep everything that comes 'before' this partition. Since the matter of a Number is an ordinal, a 'point' of partition is an element of this ordinal, and thus a smaller ordinal. What there is 'before the partition' is constituted by the ordinals smaller than the one that defines the partition. But the ordinals smaller than a given ordinal are precisely the elements of that ordinal. Consequently, if w is the point of the partition, then what comes before it, being constituted by all the elements of w, is nothing other than w itself. By partitioning at point w we obtain a new ordinal–matter, which is w – a matter evidently more limited than that from which it was cut.

But, it will be asked, what happens to the form, the numerical section from the matter? Here, once again, the idea is very simple: as the form of the new Number, we keep precisely those ordinals that are in the form of the partitioned Number and which are 'before' the partition. A sub-Number will truly be a segment of a Number, up to point w, retaining up to w (that is, between 0 and w) all the characteristics of the partitioned Number.

Let's give all of this a more precise form. Take a Number $N_1 = (W_1, F(N_1))$ and an ordinal w which is an element of W_1 (i.e. is in the matter of N_1). We partition N_1 at point w, retaining only ordinals that are lower than w, without changing the rest at all: the elements of the form of the new Number will be those of $F(N_1)$ *that are lower than* w. We thus make use of a property possessed by every set of ordinals (and therefore by the form of every Number) (see N3): because it is composed of ordinals, its elements are ordered by the relation $\in$. It is therefore entirely proper to speak of 'all the ordinals of $F(N_1)$ smaller than the ordinal w'. The diagram (compare **12.4**) shows this:

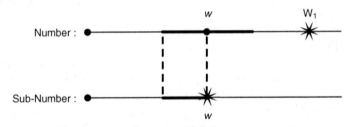

We write E/w for the segment up to point w of a set E of ordinals of which w is an element. E/w contains only elements of E lower than w (but *not* w itself, please note!). The Number obtained by the partition of N_1, and which, by extension of our notation, we will call N_1/w (which means that w must be in the matter W_1 of N_1, that $w \in W_1$), will have as its code: $(w, F(N_1)/w)$. Its matter is w – the point at which it is partitioned, an ordinal that comes 'before' W_1 – and its form is composed of all the ordinals in the form of N_1 which are smaller than w. By the same token, its residue is composed of all the ordinals smaller than w which are in the residue of N_1.

We should note that this Number $(w, F(N_1)/w)$ is exactly 'like' N_1 *up to the ordinal* w *(exclusive)*: in fact, up to w, any ordinal that is in the form of N_1 is in the form of $(w, F(N_1)/w)$ too, and an ordinal that is in the residue of the former is also in the residue of the latter. The new Number obtained through partition is, in short, the 'initial segment' of N_1, an exact copy of the 'beginning' of N_1.

Take two Numbers N_1 and N_2. If there exists an ordinal w such that $N_1 = N_2/w$, where N_1 partitions N_2 at point w, then we say that

N_1 is a *sub-Number* of N_2. Or, alternatively: a sub-Number of N_1 is a segment N_1/w of N_1.

14.5. One sub-Number of N_1 – and one only – can be defined for every ordinal w in the matter of N_1: therefore for every element of W_1. There exist exactly W_1 sub-Numbers of N_1, since an ordinal 'counts' the ordinals that precede it. Generally speaking, a Number admits of as many sub-Numbers as there are ordinals in its matter.

14.6. Take N_1/w, a sub-Number of N_1. It is clear (see the definitions and the diagram) that w is the discriminant of N_1/w and N_1, since, up to w, these two Numbers are identical. Now, the matter of N_1/w is w. So w is outside the matter of N_1/w. The order-relation between N_1 and its sub-Number N_1/w will therefore depend entirely upon the location of the ordinal w in the Number N_1: whether w is in its form or in its residue.

There are therefore *two types* of sub-Numbers for a given Number N_1:

1 Sub-Numbers N_1/w_2 where w_2 – which is at once their matter and the discriminant of themselves and N_1 – *is in the form of N_1*. These sub-Numbers are *smaller* than the Number N_1 (the discriminant w_2 is outside the matter of N_1/w_2 and in the form of N_1).
2 Sub-Numbers N_1/w_3 where w_3 *is in the residue* of N_1. These sub-Numbers are *larger* than the Number N_1 (the discriminant w_3 is in the residue of N_1 and outside the matter of N_1/w_3).

A sub-Number N_1/w_2 of the first type will be called a *low* sub-Number. A sub-Number of the second type will be called a *high* sub-Number. The following diagram shows a low sub-Number and a high sub-Number:

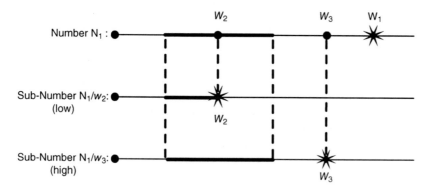

Note that there are evidently as many lows as there are elements in the form of N_1 (w_2 must be in the form), and that there are as many highs as there are elements in the residue of N_1 (w_3 must be in the residue).

The *low set* of Number N_1, denoted by $Lo(N_1)$, is the set of low sub-Numbers of N_1. The symmetrical case (the set of high sub-Numbers) is denoted by $Hi(N_1)$, to be read 'high set of Number N_1'.

14.7. The crucial point, then, is the following. Take a Number N, its low set $Lo(N)$ and its high set $Hi(N)$. N is *the one unique Number of minimal matter* to be situated 'between' the sets of Numbers which are its high and its low sets.

This can be stated precisely as follows:

1   N is situated 'between' $Lo(N)$ and $Hi(N)$ in the sense that it is larger than all the Numbers of one and smaller than all the Numbers of the other.
2   All the other Numbers situated between $Lo(N)$ and $Hi(N)$ have a greater matter than those of N. N is therefore the only Number of minimal matter to occupy the interval between its low set and its high set. Thus a Number N is a 'cut' between its low set and its high set, a cut defined 'up to matter'.[3] The two sets of sub-Numbers $Lo(N)$ and $Hi(N)$ define N itself by way of location (between the two) and material minimality.

14.8. The statement that N is between its high set and its low set is quite trivial, since by definition all the low sub-Numbers are smaller than N and all the high sub-Numbers are larger than N. The problem is to establish that N is of minimal matter between the Numbers thus situated, and that it is the only one to have this matter.

14.9. Principal lemma
Take N_1, a Number, and N_2, another Number, smaller than N_1 and of lesser matter than N_1 (so that $N_2 < N_1$ and $M(N_2) < M(N_1)$). Then *either* N_2 is a Number from the low set of N_1, *or* there exists a Number from the low set of N_1 situated between N_2 and N_1.

Let w be the discriminant of N_1 and N_2. Since we suppose the matter of N_2 to be lower than that of N_1, and since $N_2 < N_1$, w is necessarily in the form of N_1 (it cannot be in the residue of N_2 and outside the matter of N_1, because then it would be in the matter of N_2 and outside the matter of N_1, which possibility is excluded by the fact that $M(N_2) < M(N_1)$). Consider the sub-Number N_1/w. Since w

is in the form of N_1, it is a sub-Number from the low set of N_1 (it is smaller than N_1).

Up to, but excluding, w, N_2 and N_1 are identical. If the discriminant w is outside the matter of N_2 and therefore equal to its matter, N_2 is none other than the sub-Number N_1/w, and is therefore a sub-Number from the low set of N_1. If w is in the residue of N_2, then N_2 is smaller than the sub-Number N_1/w, because the discriminant of N_2 and N_1/w is necessarily w – N_2 being identical to N_1 up to the ordinal w (exclusive), and therefore also identical to N_1/w, which is a partition at w of N_1, up to w (exclusive). Now, w is outside the matter of N_1/w, so we must suppose that it is in the residue of N_2. So $N_2 < N_1/w$.

Thus it is established that N_2 is indeed either a Number from the low set of N_1 or smaller than a Number from the low set of N_1.

14.10. An absolutely symmetrical chain of reasoning would prove that, if $N_1 < N_2$ and N_2 is of a lesser matter than N_1, then either N_2 is a Number from the high set of N_1, or else there exists a Number from the high set of N_1 situated between N_1 and N_2.

14.11. Conclusion: for every number lower than (or, respectively, higher than) N_1 and of lesser matter than N_1, it is the case either that it is a Number from the low set (or, respectively, the high set) of N_1, or else that a Number from the low set (or high set) can be intercalated between it and N_1. It is therefore impossible for any of these numbers to be situated 'between' Lo(N_1) and Hi(N_1) (to be higher than every element of Lo and lower than every element of Hi) whilst at the same time being of lesser matter than N_1. The result is that N_1, which is indeed situated between its low set and its high set, is of minimal matter with regard to all Numbers thus situated.

14.12. We will now demonstrate that N_1 is *the only* Number of minimal matter situated between its low set and its high set.

Suppose there existed another Number N_2, situated between the low set and high set of N_1, and of the same matter as N_1. Such a Number could be represented as follows (with some abuse of our notation):

$$\text{Lo}(N_1) < N_2 < N_1 < \text{Hi}(N_1)$$

Since N_2 is of the same matter as N_1, the discriminant w of N_1 and N_2 is necessarily in the residue of N_2 and in the form of N_1. This means that the sub-Number N_1/w is in the low set of N_1. Now this sub-Number, N_1/w, is manifestly larger than N_2 (their discriminant, once again, is w, which is in the residue of N_2 and outside the matter of N_1/w). Thus it cannot be the case that N_2 is larger than *every* Number in the low set of N_1.

If we had the arrangement:

$$Lo(N_1) < N_1 < N_2 < Hi(N_1)$$

– we could demonstrate in the same way that there must exist a Number from the high set of N_1 which is smaller than N_2 (a good exercise).

It follows that N_1 really is the only Number of minimal matter to be situated between $Lo(N_1)$ and $Hi(N_1)$.

N_1 is *identified*, 'up to matter' – as the unique minimal element of that matter, once the 'between' position has been fixed – by the cut of two sets of Numbers, the low set and the high set. We shall write: $N_1 = Lo(N_1)/Hi(N_1)$. We shall call the cut $Lo(N_1)/Hi(N_1)$ the *canonical presentation* of N_1.

14.13. A remarkable characteristic of the canonical representation of N_1 is that all the elements of Lo and of Hi are sub-Numbers of N_1. Every number can be represented on the basis of Numbers deducted from lesser matters than their own.

The canonical presentation is a framing[4] of Number from above and below, realised by means of more tightly controlled sections than those carried out by Number.

Every Number is a cut within sets of sub-Numbers, every Number operates at the limit of two series of Numbers subordinate and immanent to it.

With this, the structuralisation of the concept of Number is complete. Not only can a Number be located as a section cut from natural multiplicities, but this section can itself be presented as a point of cutting between two series of sections of the same type. A Number is precisely thinkable as the hinge of its sub-Numbers. Number, so far from being a simple entity, answers to theorems of decomposition: it is a structure localisable in thought as a point of articulation of its substructures.

A Number exhibits, as a one-result, its immanent numerical determinations.

15

Cuts: The Fundamental Theorem

15.1. And so, let us penetrate into the swarming of Numbers.

A first remark, concerning what might be called the number of Numbers: this number is precisely not a Number, it is not even a consistent multiplicity. Numbers are numberless.

In fact, given that a Number is the pair of an ordinal and of a part of that ordinal, not only are there at least as many different Numbers as there are different ordinals, but there are *many more*, even if this 'more' flickers beyond the frontiers of sense. For *each* ordinal, there are as many different Numbers as there are different parts of that ordinal: if W is an ordinal, serving as the matter of certain Numbers, there will be $p(W)$ (the set of parts of W) forms – each one virtually extractable by means of a numerical section from this matter.

Now we already know that the ordinals do not constitute a set. 'All' the ordinals cannot be counted for one in a set-theoretical recollection. In other words, the ordinals form an inconsistent multiplicity. Consequently, the same goes for Numbers.

But, what is more, for any given multiple whatsoever, we *cannot know* exactly what the quantity of the set of its parts is. Certainly, we know (Cantor's theorem) that it must be larger than that of the initial set: it is always the case that $p(W) > W$. But 'how much' larger? It has been proven (by Gödel and Cohen's theorems) that the amount of this excess is undecidable on the basis of the fundamental axioms of set theory. In fact it is coherent within these axioms to say that $p(W)$ is 'immensely' larger than W; and it is also coherent to say that it is 'minimally' larger.[1]

Ultimately, for every ordinal, there are always *more* possible Numbers for which it is the matter than there are elements of the ordinal itself. And, as things currently stand, the extent of this 'more' can only be *decided*. Which is to say that the number of Numbers is an inconsistency of inconsistencies.

The simplest way to put it is to say: *Number is coextensive with Being*. It inconsists, is disseminated and profused just like the pure multiple, the general form of being qua being.

15.2. This inconsistent swarming of Numbers gives us to anticipate the difficulties that arise with regard to the *identification* of a specific Number picked 'from the crowd'. Every Number is cemented into the throng of those that pack in tightly, on its right (Numbers larger than it) and on its left (smaller Numbers). No Number simply, uncomplicatedly succeeds any other. Every microzone of the numerical domain teems with a numberless horde of Numbers. The numerical topology is peculiarly *dense*. And this is the problem: is it possible to identify *a* Number as opposed to sets of Numbers? Or must we consign ourselves, when we consider series of Numbers, infinite sets of Numbers, to being unable to attach to them, univocally, any specific Number? Does the numberless throng of Numbers necessarily lead us into 'those indefinite regions of the swell where all reality is dissolved'?[2]

This is where trans-numeric inconsistency summons us to think the *cut*. Is it possible, in a fabric so dense that nothing any longer numbers it, to cut *at a specific point*? Can one determine, by cutting, a *singular* Number?

15.3. This problem is not in the least bit academic, nor is it relevant solely to the thinking of Number. We are told every day how 'the complexity of modern society' prevents us from making any cut, any intervention. Contemporary conservatism no longer argues from the sacredness of the established order, but from its density. Every local cut, it says, is really a 'tear in the social fabric'. Leave natural laws (the market, appetite, domination) to operate – because it is impossible to interrupt them at any point. Every point is too dependent on all the others to permit the precision of an interrupting cut.

Thinking the cut in the hyper-dense, closely knitted fabric of Numbers will allow us to conclude that such arguments are fallacious. Every point *separates* dense sets of Numbers, every Number is the place of a cut, and, conversely, every cut prescribes one Number and one only. Not 'indefinite regions', but '*a* Constellation'.[3]

15.4. This problem also has a complex philosophical genealogy: that of the dialectic between *continuous* magnitude and *discrete* magnitude. If the being of the continuum is grasped in its intimate coalescence, so that it is not constituted from distinguishable points, but rather from complicated 'neighbourhoods', it must be thought as disjoint from discrete quantity, which enumerates successive marks. Up to, and including, Hegel, this opposition, which subsumes and underwrites that between geometry and arithmetic, remains in the position of an enigmatic real for the philosophy of quantity. In Kant, still, it ultimately supports the duality of forms of sensibility: Space is the transcendental figure of the continuous, Time – from which proceeds number – that of discrete succession.

The most profound concept of the cut, a concept that plays an immense role in modern thought,[4] displaces and refounds the dialectical schema which considers the couplet discrete/continuous to be the founding contradiction of the quantitative. This concept brings forth a singularity – and therefore a basis for distinction – in the fabric of the continuous, in the dense stuff of infinitely small neighbourhoods. Overturning the customary order of thought, it shows how a certain sort of interruption of the continuum defines a type of discreteness. Rather than saying that the continuum is composed of points, it determines points within the continuum, and even *defines* punctuality on the basis of a cut in the continuum. The concept of cut substitutes, for a problematic of *composition*, a problematic of *completion*: a point comes to 'fill in' a juncture, or an imperceptible lacuna, in a pre-given continuity.

15.5. Dedekind[5] invented the concept of the cut in order to define irrational numbers.

He begins with rational numbers. We know that a positive rational number is of the form $\frac{p}{q}$, p and q being natural whole numbers. The rational numbers provide our primary image of continuity owing to the fact that their order is *dense*. A dense order is an order such that between two ordered elements is always intercalated a third – and, by reiteration of this property, an infinity of elements. If we take the rational number 0 (which is rational because it can also be expressed as any fraction $\frac{0}{p}$) and the rational number $\frac{1}{2}$, then $0 < \frac{1}{2}$. But the numbers $\frac{1}{3}$, $\frac{1}{4}$, $\frac{1}{5}$, etc. – and an infinity of numbers of the form $\frac{1}{n}$ – intercalate themselves between 0 and $\frac{1}{2}$.

Density does not directly express a quantitative property: the rational numbers are an infinity of the type belonging to the countable, an infinity no greater than that of the natural whole numbers, and the latter, being none other than the finite ordinals, do not present a dense order: there is no natural whole number between n and $n + 1$. Density is really a topological property *of order*: excluding the simple idea of 'another step', of the well-determined *follower* of a term, it proposes instead a sort of general coalescence in which every term 'sticks' to an infinity of neighbours. The density of an order is a topological property, whereas succession is an algebraic property. Density is 'quasi-continuous', one can approach a rational number as closely as one wishes through other rationals. One even gets the feeling that, between two rational numbers and, more generally, between two terms of a dense order, there is no place for numbers, or terms, *of another type*, since the whole interval, no matter how small it is, is already populated with an infinity of rationals, or an infinity of terms of the dense order.

Now, it is precisely in this quasi-continuity of rationals that Dedekind will, by means of the cut, define additional 'points' that will complete the apparently uncompletable density of the rationals and obtain a 'true' continuum, through interruptions in their quasi-continuity.

We will return in greater detail to this procedure in chapter 16. But schematically: Dedekind considers disjoint sets of rational numbers R_1 and R_2, for which every element of R_1 is less than every element of R_2 and which, R_2 having no rational internal maximum nor R_1 any rational internal minimum, are two 'open' sets, one high, the other low. Dedekind then *identifies* a real number as occupying the place of a cut between R_1 and R_2. This real number will be both the upper limit of R_1 and the lower limit of R_2. The density of the order of rationals plays an essential role in this construction, once it is understood that density and the cut, far from being exclusive, are paired together in thought.

It must be noted straightaway that this procedure seeks to *define* real numbers, the rational numbers being supposed to be known. The Dedekind cut is wholly an operation of completion: where there is *nothing*, no rational number, the name of something 'extra' comes forth. The real number defined by the cut R_1/R_2 fills in that which, thought purely from the point of view of rationals, is *a void in the density*, and thus a void to which nothing attests. This is why the cut *founds* a new species of numbers, which 'complete' the initial density and retroactively indicate that this density was not so dense that gaps could not be discovered therein.

15.6. We cannot hope to 'complete' the inconsistent domain of Numbers, nor to found, outside Number, a hyper-number which would name the invisible lacunae in it. Our Numbers are uncompletable, being coextensive with Being (see **15.1**). *All* the Numbers are already there. What could a cut mean in such conditions?

Nevertheless, there is a very strong concept of the cut for Numbers. This concept holds 'up to matter', like that of the singular element separative of a Number and its sub-Numbers, of its identity as cut between its low set and its high set (see chapter 14).

This concept of the cut is presented in the following theorem which, articulating the inconsistent swarming of Numbers with the precision and the unity of a punctual cut, well deserves the name of *fundamental theorem of the ontology of Number*:

> Given two sets of Numbers, denoted by B (for 'from below') and A (for 'from above'), such that every Number of set B is smaller than every Number of set A (in the order of Numbers, of course), **there always exists one unique Number N of minimal matter situated 'between' B and A.** 'Situated between' means that N is larger than every element of B and smaller than every element of A.

The Number N is evidently not the only one between B and A. The numerical swarming is such, the density is so considerable, that such a solitude would be unthinkable. But it is the only number to be found *with its matter*. All the others have a larger matter, in a rigorous sense, since matters are ordinals: the ordinal–matter of N is minimal for the property 'is the matter of a Number situated between the sets of Numbers B and A'.

It will not surprise us at all to find minimality here: it is a classic organisational principle of ordinals. What is surprising is:

– that such a Number should *exist*;
– that it should be *unique*.

Its existence founds the principle of the cut. If two sets of Numbers are like B and A (every Number of B being smaller than every Number of A), then one can still speak of what exists 'between' B and A and is neither of B nor of A, in spite of the prodigious density of the order of Numbers. It is thus possible to make a *cut* in the hyper-dense fabric of this order.

Uniqueness (up to matter, which is to say uniqueness of the Number–cut of minimal matter) founds the principle of identification, the persistence of the count-for-one even where all is coalescent, in dense neighborhoods. A cut designates *one* Number, and designates it on the basis of sets of Numbers. We will hold that no complexity, even one pushed to the point of inconsistency, no density, even one pushed to the finest infinitesimal proximities, can authorise the prohibition against cutting at a point.

15.7. The rest of this chapter is dedicated to the demonstration of the fundamental theorem, the only theorem in this book that is a little complex.[6]

I do not, moreover, intend to give all the details of the proof. However, we are at the heart of the mathematics of Number, and what must be put into play in order to think the cut is of a conceptual interest far surpassing mathematical ontology. All truth-procedures proceed via a cut, and here we have the abstract model of *every* strategy of cutting. The intellectual effort demanded of the reader will lead him or her, I am quite sure, to beatitude in the Spinozist sense.

15.8. Upper bound of a set of Numbers

Since we are engaged in investigations whose character is topological, and since in particular we are wondering how to find Numbers larger (or, respectively, smaller) than a given *set* of Numbers, let's begin with the simplest concept, that of an upper bound: given a set of Numbers, does it make sense to speak of a 'unique' Number larger that all those in the set?

Once more we must, in view of the proliferation of Numbers, avail ourselves of a concept 'up to matter'. We will prove the following: if B is a set of Numbers, then there exists a Number N which is the unique Number of minimal matter to be larger than all the Numbers in set B. We will call this N the upper bound of B. Right away the upper bound exhibits a surprising characteristic: it is *always* a Number written (W,W) – that is, a Number whose form is its whole matter.[7]

15.9. Take B, a set of Numbers. Consider the ordinal defined as follows: 'the smallest ordinal W such that, for *every* Number N of set B, there exists a $w_1 \in W$ which is either in the residue or outside the matter of N'.

One such ordinal W exists, *because B is a set, and is therefore consistent.* If W did not exist, that would mean that *all* ordinals would fall into the form of at least one Number N of B. But 'all ordinals' is an inconsistent multiplicity, and consequently B would also be an inconsistent multiplicity, and would not be able to be thought as a set.

That there should exist such a 'smallest' W results from the property of minimality that characterises the ordinals.

W having been specified, now consider the Number (W,W). This number is larger than every Number in set B. In fact, by the definition of W, for every Number N of B there exists a $w_1 \in$ W which is in its residue or outside its matter. Now, as the form of (W,W) is W, every $w_1 \in$ W is in the form of (W,W). The discriminant of a Number N and of (W,W) is necessarily the smallest $w_1 \in$ W that is in the residue or outside the matter of N. And, since this w_1 is in the form of (W,W), the residue (or outside-matter)/form relation demands that (W,W) should be larger than N.

Since a Number larger than every Number in B exists – namely (W,W) – one of minimal matter must exist, in virtue of the ordinals' property of minimality.

Therefore, let (W_1,X) be a Number of minimal matter for the property 'being the matter of a Number larger than all the Numbers in B'. Its form X is in fact equal to W_1.

For, if X differed from W_1 – if, that is, the form of the Number was not its whole matter – that would mean that there existed at least one ordinal $w_2 \in$ W which was in the residue. Consider then the sub-Number of (W_1,X) obtained by partition at w_2 – that is, the sub-Number $(w_2,X/w_2)$. Since w_2 is in the residue of (W_1,X), the sub-Number $(w_2,X/w_2)$ is in the high set of (W_1,X) (see **14.7**). It is therefore larger than (W_1,X), and a fortiori larger than every Number in B, since this is already the case for (W_1,X).

But that is impossible, because the number $(w_2,X/w_2)$ is of lesser matter than the Number (W_1,X). Now, we supposed that (W_1,X) was of minimal matter for Numbers higher than every number in B.

Our initial hypothesis must be rejected: there does not exist in (W_1,X) any element that is in the residue, which is to say that the form occupies the whole matter, and that the Number must be written (W_1,W_1).

There exists therefore *one Number only* of minimal matter that is higher than all the Numbers of set B: it is the Number (W_1,W_1), where W_1 is this minimal matter.

We can thus legitimately speak of *the* upper bound of a set of Numbers. Already the theme of unicity comes to inscribe itself as bar, or caesura, in the hyper-dense swarming of Numbers.

15.10. Lower bound of a set of Numbers

Reasoning totally symmetrical with that employed for the upper bound will permit us to define the unique Number of minimal matter that is smaller than a set A of Numbers. This will be the lower bound of the set A. We will see that, this time, this Number is written $(W_2,0)$:

its form is void, the numerical section does not extract anything from the matter W_2.[8]

Let A be a set of Numbers, and let W be the ordinal defined thus: 'the smallest ordinal such that, for *every* Number N of A, there exists a $w_1 \in W$ which is either in the form, or outside the matter of N'. This ordinal exists necessarily, because A is a set, and in virtue of the principle of minimality (see above).

The Number (W,0), whose matter is W and whose form is the void, is smaller than every Number of A. In fact, all ordinals $w_1 \in W$ are in the residue of (W,0). Now, for every Number N of A, by the definition of W, there exists a $w_1 \in W$ which is either in the form of N, or outside the matter of N. The smallest such w_1 is the discriminant of N and of (W,0), and its location means that (W,0) < N.

Hence there exists a Number smaller than every Number in A, and – by the principle of minimality – there exists at least one of minimal matter, say (W_2,X).

It is easy to prove that X is necessarily the empty set. If it were not, that would mean that there existed a $w_3 \in W_2$ which was in the form of (W_2,X). But then the sub-Number of (W_2,X) obtained by partition at w_3, that is, $(w_3,X/w_3)$, would be in the low set of (W_2,X) (see **14.7**). It would then be smaller than (W_2,X), and therefore smaller than every Number in A, although of lesser matter than (W_2,X): which is impossible in view of the minimality of W_2 for this position.

Therefore, there exists one unique Number of minimal matter that is smaller than every Number in A. It is the Number $(W_2,0)$, where W_2 is this minimal matter. The Number $(W_2,0)$ is the lower bound (up to matter) of set A.

15.11. Fundamental theorem, first part: Existence

'Existence' means here: existence of at least one Number situated between two sets of Numbers B and A, which, in an abuse of our usual notation, we shall generally write as B < N < A.

Take B and A, two sets of Numbers such that every Number of B is smaller than every Number of A. Our technique will consist in constructing, between B and A, step by step – that is to say, ordinal by ordinal – starting from 0, a Number N 'suspended' at every step in such a way as to assure us that nothing up to the ordinal W in question, which is to say, for every step taken in the procedure – can force the Number N to be smaller than a Number of B, or larger than a Number of A. We might also say that we are going to construct

N from its sub-Numbers of intersecting matter, by 'choosing' to put an ordinal W in the form or in the residue of the Number N under construction, depending on the relationship between the segment of the procedure of N which goes from 0 to W, and the various sub-Numbers in B and in A.

The underlying idea is that the construction of a cut makes necessary a *local* domination of the substructures implicated in the course of this construction. This is a general law of practice, at least in so far as the latter aims at effects of cutting (foundational interruptions).

This technique boasts the very great interest of highlighting the link between cutting and a sort of *procedure of neutralisation*. So that N can slide in between the Numbers of B and the Numbers of A, we are going to remain mindful of the fact that the principle of order, at every point of N, 'neutralises' the discrimination between the Numbers of B and the Numbers of A. The great difficulty being to know when to stop ourselves, when to *fix* the matter of the Number N, which we would have traversed, all the while postponing its closure.

In all domains of thought, to proceed with a precise cut in a densely ordered fabric is to calculate a prudent tactics of insertion step by step, and then to risk a stopping point which will irreversibly fix the intermediary term. The cut thus combines the neutrality of the interval and the abruptness of the interruption. This is why great strategies of thought must always attain a mastery both of the patience which, point by point, opens and enlarges a lacuna, and of the impatience which comes to seal and to name its existence from this moment forward, without return or recourse.

15.12. So we begin from the ordinal 0, and we traverse the ordinals, assigning to each a value $f(W)$ – the values being F (for form), R (for residue), or M (for matter). The value M can obviously only be given once, and last of all, because the Number N that we want to construct has only one matter. For an ordinal W, if $f(W) = F$, we will put W in the form of the Number N under construction; if $f(W) = R$, we will put it in the residue. So long as we have not assigned the value M, the sub-Numbers are still 'under construction'. The procedure amounts to fixing a location-status for each ordinal W, so that the sub-Number N/W, as the procedure continues, will appear retroactively as never constraining N to be larger than any Number of A, or smaller than any Number of B.

The strategic patience of the construction of a cut consists in inserting additional *local* values without compromising the chances of a *global* cut. It is a work that proceeds point by point, but is retroactively decided as an irreversible and general caesura.

We will denote by N*b* and N*a*, with indices if need be, the Numbers of B and A. N*i* will designate the intervallic Number, the Number we wish to construct between B and A.

If, for a given Number N*b* (or respectively N*a*), the values attributed by *f* to ordinals smaller than an ordinal W (values of the type F or R, which the ordinals of the N*i* under construction take) are exactly those which locate these ordinals in N*b* (or respectively in N*a*), then we say that W *identifies* N*i* and N*b* (or respectively N*a*). W's identifying N*b* (or N*a*) and N*i* means that, in every case, no ordinal smaller than W can discriminate between N*b* (or N*a*) and N*i*. In particular, the discriminant of N*b*, or N*a*, and the segment of N*i* under construction (a segment which ranges from 0 to W exclusive) cannot figure in the ordinals inferior to W. Which amounts to saying – and this is the most tractable form of the relation of identification at ordinal point W – that, up to W, the 'sub-Number' N*i*/W is identical to the sub-Number N*b*/W (respectively N*a*/W).

We will denote by Id.(W,N*b*) the fact that W identifies N*i* and N*b*. And the same thing for N*a*. All the while we should keep in mind that Id.(W,N*b*) means that N*i*/W = N*b*/W.

The strategic idea is to construct an N*i* 'neutralised' for order, by making sure, each time one comes to 'the end' of a series of ordinals which identify N*b* (or respectively N*a*) and the N*i* under construction, that the choice of a value for *f*(W) will not be able to compromise our chances of positing a hypothetically completed N*i*, which would be intervallic between B and A. We must just make sure that no ordinal comes to be in the position of an unfavourable discriminant forcing N*i* to be smaller than a Number of B, or larger than one of A. The prudence of the cut consists here in never risking losing the chance to take up an intervallic position. Conserve its chances, that is the maxim of the 'step by step' phase of the construction of a cut.

15.13. We will posit the following rules – rules of construction of N*i* for the ordinals starting from 0:

RULE 1: If Id.(W,N*b*) and W is the matter of N*b*, then *f*(W) = F. We put the ordinal W in the form of N*i* whenever, at the end of an N*b*/W identical to N*i*/W, W is the matter of N*b*. So, using a black square to denote a belonging to the form:

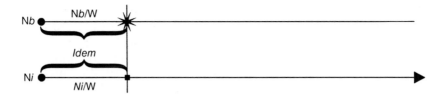

RULE 2: If Id.(W,Na) and W is the matter of Na, then $f(W) = R$. The diagram should be clear, marking with / a belonging to the residue:

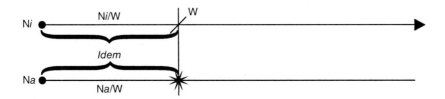

RULE 3: If rules 1 and 2 do not apply for a given W, but instead we have an Nb such that Id.(W,Nb) with W in the form of Nb, then $f(W) = F$. If cases 1 and 2 do not apply, we put W in the form of Ni each time that, at the end of an Nb/W identical to Ni/W, W is in the form of Nb:

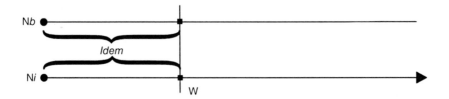

RULE 4: If rules 1 and 2 do not apply, and we have an Na such that Id.(W,Na) with W in the residue of Na, then $f(W) = R$:

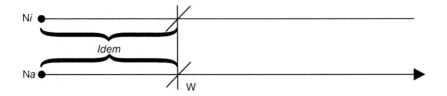

RULE 5: If none of the first four rules apply, it must be the case that, for the W considered, no Nb such that Id.(W,Nb) has W for matter or in its form, and that no Na such that Id.(W,Na) has W for matter or in its residue. Under such conditions, at point W, if there exists an Nb for which Id.(W,Nb), W is in the residue of Nb, and if it is the case that Id.(W,Na), W is in the form of Na. We then say that f(W) = M, which completes the construction of Ni.

As justification for this rule, note the following: since all Nb (or respectively Na) where W *is not* thus located – so, W in the residue (or respectively in the form) – are such that W does not identify them with Ni, then, for these Nb (or Na), it is the case that Nb/W ≠ Ni/W (or respectively Na/W ≠ Ni/W). In other words, these Nb and these Na have *already* been discriminated, before ordinal W, by the process Ni. The only Nb and Na not to have already been discriminated are those where W is in the residue (or respectively in the form).

Given this remark, we can state that *rule 5 prescribes with complete justification the decision of closure of the process* Ni. We can posit: f(W) = M, thereby fixing W as the ordinal–matter of Ni, and therefore as that place where the process of the construction of Ni ends.

If W is the matter of Ni, it is located outside the matter for that Ni *supposed closed in* W. Now, W does not discriminate Ni from Nb where W is in the residue, or from Na where W is in the form. The location for Ni will remain 'between' B and A, since the schema of the order-relation is precisely R < oM < F. We will have:

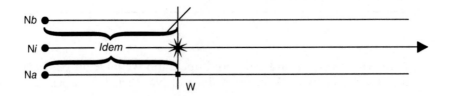

Closure is entirely possible, since, beyond ordinal W, *all* Nb and Na are discriminated by Ni (*before* W through rules 1 to 4, *at point* W by rule 5); and our rules reflect the fact that this discrimination always goes in the direction Nb < Ni < Na.

This regulation, however, merits immediate examination.

15.14. It is essential to confirm that our rules do not contradict one another.

Take for example rules 1 and 2. If by some mischance it should happen that *at the same time* Id.(W,Nb) and Id.(W,Na), with W the matter both of Nb and of Na, then W would have to be placed simultaneously in the form and in the residue of Ni . . .

But such a case cannot arise. Because, if W is the matter of Nb and of Na, since every Number of B is smaller than every Number of A, it is the case that Nb < Na. And, since they have the same matter W, their discriminant must be less than W, which is to say that there is at least one ordinal w_1 ∈ W which doesn't have the same location in Nb and in Na. It is therefore not possible for sub-Numbers Nb/W and Na/W to be identical. This means, moreover, that, if both Id.(W,Nb) and Id.(W,Ba), their common identity must be Ni/W. So rules 1 and 2 are compatible.

But take rules 3 and 4. If by some mischance there is a W for which rules 1 and 2 do not apply, and there exist Nb and Na for which, firstly, Id.(W,Nb) and Id.(W,Na), and, secondly, W is in the form of Nb and W is in the residue of Na, W would have to be placed both in the form and in the residue of Ni.

But of course such an unfortunate circumstance cannot arise. Because, if W is in the residue of Na and in the form of Nb, then it discriminates between Nb and Na. But this could not be their discriminant, otherwise it would be the case, with regard to this location, that Na < Nb, which is prohibited by B < A. Therefore the discriminant is *smaller* than W, and, as before, it is impossible that Nb/W = Na/W; which makes it necessary to suppose their common equality to Ni/W.

15.15. Now we will see whether, with these rules, we do indeed preserve our chances that Ni will slip in between *all* the Numbers of B and *all* the Numbers of A, and therefore between all Nb and all Na.

When we apply rule 1, we give the value F to the ordinal W. This certainly cannot make Ni become less than a Number of B, because, if W is the discriminant of Ni and of an Nb, being in the form of Ni, it will always be the case that Nb < Ni.

But, given the fact that we put W in its form, don't we risk Ni becoming larger than a Number of A? For this it would have to be the case that W was the discriminant of Ni and of an Na. But then it would also ultimately be the discriminant of the Nb of which W is the matter (since we apply rule 1) and of Na. Now, we know that Nb < Na. If their discriminant is the matter of Nb, it must be in the form of Na. This location of W – W being the discriminant of Ni and Na – prohibits us from having the order Na < Ni.

So, in applying rule 1, we can be sure that the location that we fix for W in the Number Ni under construction *entails neither an unwelcome and frustrating* Ni < Nb, *nor a fatal* Na < Ni. At point W, Ni stays situated 'between' B and A.

The examination of the other rules leads us to the same conclusion. Let's carry out this examination for rule 5 (for rules 2, 3 and 4 the methods are the same as for rule 1. Let the reader prove this as an exercise, with the help of the note,[9] and above all of the diagram below.)

Rule 5 comes into play when rules 1 to 4 are not applicable. The W under consideration makes no identification between any Nb (or Na) and Ni if W is located as matter of Nb (rule 1), matter of Na (rule 2), form of Nb (rule 3) or residue of Na (rule 4). If, then, it is the case that Id.(W,Nb), or Id.(W,Na), it is because W is in the residue of Nb and/or in the form of Na. These two hypotheses are compatible this time: the identifications in question could obtain, and W could be *both* in the residue of Nb and in the form of Na. Rule 5 then compels us to make the gesture of closure $f(W) = M$, which determines W as matter of the intervallic Number Ni. In the Ni thus closed, W is located outside the matter. Can this choice make Ni less than some Na, according to the relation R < oM? No, because, if W discriminates between this Na and Ni, with W in the residue of Na, this would be a case for the application of rule 4, which would exclude the use of rule 5. And, in the same way, it cannot be the case that Ni < Nb according to the relation oM < F, because the location of the discriminant W in the form of an Nb compels, for W, the use of rule 3 rather than rule 5. Rule 5, applied when it is proper to do so, cannot entail that Ni < Nb. And, as it cannot entail Na < Ni either, it leaves the procedure Ni, at point W, in the interval between B and A.

So it is that, at every ordinal point W, the application of our rules 'locally' situates Ni, in the form of the sub-Numbers Ni/W, in an intervallic position with regard to B and A. Our step-by-step labour is pursued without Ni surpassing any Na, or being surpassed by any Nb. We conserve our chances all the way through the construction. An enlarged diagram shows how Ni proceeds. We have, above, some Numbers Nb of B, below, some Numbers Na of A, and, in the middle, the process of Ni. The ordinals W_1 to W_5 present, in order, cases of the application of the five rules. Squares, asterisks and bars designate form, matter and residue. You will recall that, when a point is marked in an Nb or an Na, it means that, *before* that ordinal point, Na (or

N*b*) is identical to N*i* (relation of identification at an ordinal point).

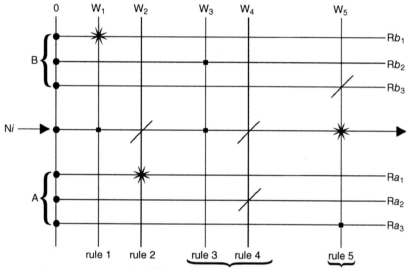

The whole subtlety of the enterprise lies in minimising the risks, in making sure not to increase the value of N*i* to point W (in particular, in not giving it value F) until one is sure that this increase will have no effect with regard to A; and in not decreasing this value (the value R) unless all effect with regard to B is excluded. Thus N*i*, perpetually maximising the neutralisation of the effects of order, slips in between B and A.

And, when the time for closure arrives (rule 5), for a W situated between residue (N*b*) and form (N*a*), we retroactively set the seal on the tactics, arriving at a Number *globally* situated between B and A, because it is protected, *locally*, from any prohibition against this possibility.

15.16. Fundamental theorem, second part: Unicity

We have just indicated the strategy – combining local, neutralising patience with a global decision of closure – that allows the existence to be established, in every case, of at least one Number situated between two sets of Numbers B and A such that (in an abuse of notation) B < A. In virtue of the principle of minimality of ordinals, there must exist at least one such Number of minimal matter: we will consider the property 'being the matter of a Number situated between B and A', and the minimal ordinal for this property.

It remains to be shown that a Number of minimal matter situated between B and A is unique, which will permit us to identify *the* numerical cut between B and A.

Suppose that there were two: we would have the following arrangement:

$$B < N_1 < N_2 < A$$

– with N_1 and N_2 being of the same matter (minimal for this location).

Since N_1 and N_2 are of the same matter, $N_1 < N_2$ means that the discriminant must be in the residue of N_1 and in the form of N_2. Take this discriminant, w. Consider the sub-Number N_1/w of N_1. Since w is in the residue of N_1, this sub-Number belongs to the high set of N_1: it is therefore larger than N_1. But, since w is the discriminant of N_1 and N_2, and therefore the smallest ordinal to discriminate between them, then N_1 and N_2 are identical up to w (exclusive). This means that the sub-Number N_1/w is identical to the sub-Number N_2/w. The discriminant of N_1/w and of N_2 can only be w, which is outside the matter of N_1/w and in the form of N_2. Consequently, $N_1/w < N_2$.

So finally, we have the arrangement:

$$B < N_1 < N_1/w < N_2 < A$$

Which is to say that N_1/w is also situated between B and A. But this is impossible, given that it is of lesser matter than that of N_1, which is supposedly minimal for the location between B and A.

We must reject our initial hypothesis: there are not two Numbers of minimal matter between B and A, there is only one.

The two sets B and A therefore determine univocally *one* Number of minimal matter situated between them. This Number will be called the cut of B and A, and we will posit that N = B/A, each time that N can be identified as the unique cut of B and of A.

15.17. There is one very peculiar case of the cut: take two Numbers N_1 and N_2 such that $N_1 < N_2$. And take, for B and A, the sets which have for elements only N_1 and only N_2, that is, the singletons (N_1) and (N_2). We remain within the parameters of the fundamental theorem, which is to say that there exists a unique Number N_3 of minimal matter situated between N_1 and N_2. We thus rediscover here the classic condition of *density* of an order, which we have mentioned with regard to the rationals: between two Numbers there always exists a third, and thus an infinity of Numbers. For us, besides this, there is an additional determination: between N_1 and N_2 there is always a *unique* Number of minimal matter.

We can therefore put forward a principle which everything gives us to expect, and of which the unicity of the cut provides the infinitely strong concept: the order of Numbers is dense.

But more profound than this is the correlation in thought between this numberless density, this coalescence which inconsists in the approach to all Number, and the possibility of counting for one the Number of minimal matter which cuts the fabric without lacuna of numericality at a certain point.

'Cut' here designates the incision of thought in the inconsistent fabric of being, that which Number sections from the ground of Nature. It is a concept of singularity. Perhaps *the* concept of singularity, at least in the order of being. For there is that other singularity which cuts across being, and which is the event.

16

The Numberless Enchantment
of the Place of Number

16.1. A review, to begin with.

1   A Number is an ordinal – the matter of the Number, M(N), in which is sectioned a part of that ordinal – the form of the Number, F(N). We also consider that part of the ordinal–matter that falls outside the section, outside the form: the residue of the Number, R(N).

2   The location of an ordinal with regard to a Number N is its position in (belonging or non-belonging to) one of the three 'components' of Number: form, residue, matter. There are three locations: in the form, in the residue and outside the matter.

3   The discriminant of two Numbers N_1 and N_2 is the smallest ordinal not to be located similarly in both Numbers. If no such discriminant exists, then the two Numbers are equal (they have the same matter, the same form, and therefore the same residue).

4   Depending on the location of the discriminant, we can define an order-relation (transitive and non-reflexive) between two different Numbers. We denote this through $N_1 < N_2$ and by saying that N_1 is smaller than N_2. This relation is a total order over the domain of Numbers in the sense that, given two different numbers N_1 and N_2, it is always the case either that $N_1 < N_2$ or $N_2 < N_1$.

5   The order-relation is dense: given two Numbers N_1 and N_2 where $N_1 < N_2$, there always exists an N_3 which comes in between N_1 and N_2: $N_1 < N_3 < N_2$.

6   Take a Number N_1 of matter W_1 and an ordinal w_1 smaller than W_1 (so that $w_1 \in W_1$). The Number of matter w_1, which is exactly like N_1 up to w_1 exclusive (the form of this Number being constituted by all ordinals smaller than w_1 that are in the form of N_1) will be called a sub-Number of N_1, a sub-Number of N_1 which is a 'cut' of N_1 at point w_1. We denote this sub-Number N_1/w_1.

7   Amongst the sub-Numbers of N_1, some are smaller than N_1 (when w_1 is in the form of N_1), others are larger than N_1 (when w_1 is in the residue of N_1). The former, gathered together, constitute the low set of N_1, denoted by $Lo(N_1)$. The latter constitute the high set of N_1, denoted by $Hi(N_1)$.

8   It can be proved that N_1 is the cut of its low set and its high set in the following way: it is the Number of minimal matter situated, according to the order of Numbers, between the low set and the high set (larger than every Number in the low set and smaller than every Number in the high set).

9   More generally, it can be shown that, given two sets of Numbers such that all those of the first set are smaller than all those of the second, there exists a unique Number N of minimal matter situated between these two sets. Taking two such sets B and A, we can say that this Number N is the cut of B and A, which is written $N = B/A$. Thus $N_1 = Lo(N_1)/Hi(N_1)$. This specified cut is called the canonical presentation of N_1.

16.2. We will now take a stroll through the borderless domain of Numbers, pointing out some of them, and in particular all those traditional species: natural whole numbers, negative whole numbers, ordinals, rationals, reals. But also so many others, which finitude and the wretchedness of our inherited practice of Number keeps from us. How negligible are numbers amongst Numbers! The being of Number exceeds in every direction that which we know how to negotiate. Our strength, however, is that we possess a way of thinking of this excess of being over thought.

16.3. Zero
There is a very distinctive Number, the Number $(0,0)$, whose matter is the void, and whose form, consequently, is also the void. This Number inscribes as numerical gesture the absence of every gesture, in default of any matter. It is absolute Zero, the Number without numericality. Of course, its ontological foundation is the empty set, the suture to being of every text, the advent of being qua being to the thinkable. There is no doubt that it is this void that we are

thinking here *as Number*. But thinking it as Number makes a difference. It is not for example the same thing, not the same Number, as it would be if the void was only in the position of matter, or only in the position of form. The number ((0),0), or (1,0), whose form is void, cannot at all be identified as the Zero of Number. Certainly, the act of sectioning it is equally null, it doesn't extract anything from its matter, but this matter subsists unaltered, constituting, in the absence of any act, the real substance of that which this gesture never even started. The only true Zero is that which subtracts itself from all numerical gesture because it has *nothing*, no material or natural multiplicity, upon which this gesture could be carried out or not carried out. Zero is thus outside all appreciation, positive or negative, of the act of numerical section. It is, very precisely, neither positive or negative. It subsists in itself, inaccessible to all evaluable action. Zero is being qua being thought as Number, from within ontology.

16.4. Since we have said, a little metaphorically, that Zero is neither positive nor negative, can we not give a precise numerical sense for these adjectives? Elementary arithmetic already introduces – to the obscure relish of every schoolchild – whole negative numbers such as −4.

Consider for example the Number N_1 whose matter is the limit ordinal ω, and whose form has only the ordinal 0 as element. Which is to say that the form is the singleton of 0, and that the number N_1 can be written: $(\omega,(0))$. If we compare this Number to Zero, that is, to $(0,0)$, we can clearly see that their discriminant is 0, which is in the form of N_1 and outside the matter of Zero (any ordinal whatsoever, including 0, is outside the matter of Zero, which has no matter). The rules of order indicate to us then that N_1 is *larger* than Zero. It makes sense to say that N_1 is *positive*.

Consider now the Number N_2, whose matter is also the limit ordinal ω, but whose form is this time the singleton of 1. This Number N_2 can be written $(\omega,(1))$. Once again, the discriminant of N_2 and Zero is 0. It can be found this time in the residue of N_2, since the form of N_2 does not contain 0 (it only contains 1), but its matter, ω, does contain it, ω being the limit collection of all the finite ordinals, including 0 of course. We can see, then, that 0, being outside the matter of Zero and in the residue of N_2, N_2, is *smaller* than Zero. So it makes sense to say that N_2 is *negative*.

16.5. Positive Numbers and negative numbers
Our examples can be generalised in the following fashion: the discriminant between Zero and any other Number whatsoever is *always*

the empty set 0. For Zero is the only number whose matter is void, and therefore the only Number where 0 is located outside the matter. For every other Number, 0 is located in the form or in the residue. And, since 0 is the smallest ordinal, it is certainly the discriminant of Zero and of every number other than Zero.

The situation is very simple, then: if any Number other than Zero has 0 in its form, then it is larger than Zero. If, on the other hand, 0 is in its residue, it is smaller than Zero, since 0 will always be outside the matter of Zero.

We will thus define positive and negative Numbers in the following way: *A Number is positive if 0 is an element of its form. It is negative if 0 is an element of its residue.*

16.6. Some significant consequences of the definition of positive and negative Numbers:

1   Since Zero is without matter, without form and without residue, 0 cannot be an element either of the form or of the residue of Zero. The description in **16.3** is thus transformed into a mathematical concept: Zero is neither positive nor negative.

2   Zero is not at all the smallest Number. It is larger than every negative Number, and negative Numbers constitute, to all appearances, a limitless, inconsistent domain. Between the negative Numbers and the positive Numbers, Zero lies at the centre of that which has no periphery.

3   Zero is not defined by extrinsic operations, it is not introduced as the 'first' term of a succession, nor as the 'neutral element' of an operation (an attribute which it possesses incidentally and secondarily). It is characterised *by its numerical being*. We have not strayed from our ontological path, which subordinates all operational or algebraic considerations to immanent characterisation.

4   More generally speaking, the categories 'positive' and 'negative' have been introduced into the consideration of the order of Numbers only for convenience of exposition. The predicate 'has 0 in its form' or 'has 0 in its residue' are wholly intrinsic. The examination of the being of a Number alone tells us whether it is positive or negative, without comparing it with any other Number.

5   Positivity does not depend in the least upon the 'quantity' of the matter of a Number, or the size of its form, but only upon the location of the void. The Number $(2,(0))$ is positive, whilst the Number $(\omega,(\omega - 0))$, whose matter is ω and whose form takes in all of this matter apart from 0, is negative. There

is finite positive numericality, and infinite negative numericality, regardless of whether the question is one of matter or one of form.

6   If a Number N is positive, then, since 0 is in its form and is necessarily minimal, it follows that *every* sub-Number N/w of N (except for Zero, which is a sub-Number of every Number, the sub-Number N/0) is also positive: the elements of the form of N/w are actually the elements of the form of N up to the ordinal w, and, unless w is 0, 0 will be amongst these elements, since N is positive. Similarly, every sub-Number of a negative Number N, apart from 0, is negative (it has 0 in its residue, as N does). In particular, the non-null elements of the low set and all the elements of the high set of a positive Number are positive; likewise, all the elements of the low set and all the non-null elements of the high set of a negative Number are negative.

16.7. Meditation on the negative

The concept of negativity, as proposed by the universe of Numbers, is every bit as profound as its apparent paradoxicality suggests. One might think at first that negativity consisted precisely in the incorporation of the void into the form of Number. Isn't there more positivity in a form that has not been marked by the stigma of nothingness? Isn't the *plenitude* of the numerical section better assured if it expels from its positive production that dubious index of the multiple that allows no presentation?

Number enjoins us here to disabuse ourselves of any remaining temptation towards an ontology of Presence. If the lack of void in the form of Number seems 'positive', this is the case only if we identify being with the plenitude of the effectively presented. We are then tempted to index to the negative every occurrence of that which presents nothing, every mark whose multiple–referent is subtracted. But the truth is entirely otherwise: it is precisely under this mark that being qua being comes to thought. In which case there is *less* ontological dignity in a Number that does not retain this mark in its form than in a Number that does so retain it. It is from the point of the void that the dignity of being, the superiority of a Number, can legitimately be measured. Numerical superiority is the symbol of this superiority with regard to what is at the disposal of thought.

The ontological clarity (for a subtractive ontology) of the statement 'a Number is negative if the mark of the void is in its residue' underlies what might be called the ethical verdict of Number. I hope to show one day that what is Evil, in any situation where the void is attested to (and such, singularly, are post-evental situations), is the treating of that testimony precisely as if it were a residue of the situation. What is Evil is to take the void, which is the very being of the situation, for *unformed*. The forms of Evil declare substance full and luminous, they expel every mark of the void, they rusticate, deport, chase off, exterminate those marks. But the verdict of Number tells us: it is in this claim to full substance, in this persecution of the occurrences of the void, that resides, precisely, the negative. A *contrario*, positivity assembles and harbours the marking of the void within its forms. And, this being so, it accords thought to being in an intrinsically superior fashion.

To take the void for a residue is a negative operation, a detestable 'purification'. Every true politics, in fidelity to some popular event, takes on the guardianship of the void – of that which is unpresented, not counted, in the situation – as its highest duty in thought and in action. Every poem seeks to uncover and to carry to the formal limits of language the latent void of sensible referents. Every science treats positively the residue of its own history, that which has been left outside of its form, because it knows that precisely there dwells that which will refound and reformulate its system of statements. All love ultimately establishes itself in the joy of the empty space of the Two of the sexes which it founds, and from this point of view the romantic idea of a full, fusional love, under the purified sign of the One, is precisely the Evil of love.

The negative, as its concept is established by Number, is a *punctual* discord of thought and of being. 'Negative' is every enterprise of formation which abandons, fails to cherish, this unique point upon whose basis there can be forms and the unformed, forms and residues; the point where being, in the guise of the unpresented, assures us that we do not think in vain.

16.8. The symmetric counterpart of a Number

Not much needs to be done in order to 'negativise' a positive Number: it suffices to remove 0 from its form. Number teaches us the precarity of the positive, its a-substantial character. It is at the mercy of the transfer *of one single point* to the residue. And this point is the most transparent of all, that point that is not supported by any multiple–presentation: the mark of the void.

This idea of the transfer of a term from one location (here, the form) to the 'opposite' location (here, the residue) can be generalised. Take a Number N and the Number obtained *by inverting the form and the residue of N*: The residue of N is promoted into the form, whilst all the terms of its form are demoted into the residue. This new Number operates, in the same ordinal–matter, a cut inverse or symmetrical to that which defines N. We will call this Number the *symmetric counterpart* of N (indicating a symmetry whose centre, as we shall see, is Zero). We will denote by –N, and read as 'minus N', the symmetric counterpart of N.

A Number and its symmetric counterpart can be presented as follows (using the diagrams introduced in **12.3**):

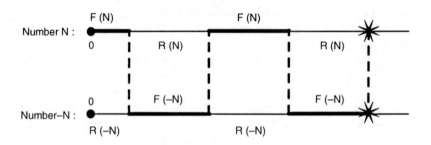

It is clear in the diagram that N is positive (0 is in its form) and that –N, its symmetric counterpart, is negative. Evidently, this will always be the case. Conversely, when N is negative (0 is in its residue), –N is positive (0 is in its form).

If we take the symmetric counterpart –N of N, then the symmetric counterpart –(–N) of –N, we arrive back at N: we have changed the form into the residue, and then the residue into the form. It is that old law learnt in the schoolroom, which spontaneously opposes itself both to Hegel and to intuitionism: two negativising operations take us back to the initial affirmation. However, one must still take care, as always, to note that –(–N) is not necessarily a positive Number. If the starting Number N is negative, its symmetric counterpart is positive, and the symmetric counterpart of its symmetric counterpart – which is itself – is once again negative. The sign '–' is not a sign of negation but one of symmetry. Which confirms for us that the negative (unlike the symmetrical) is not an operational dimension. It is a structural predicate of the being of Number.

16.9. A few examples.

What is the symmetric counterpart of the positive Number $(\omega,(0))$? It is the Number $(\omega,(\omega - (0)))$, whose form is all of ω except for 0. It is obviously negative.

What is the symmetric counterpart of the negative Number $(2,(1))$, whose form is the singleton of 1? It is the positive Number $(2,(0))$, whose form is the singleton of (0). In fact, the only elements of the ordinal 2 are 0 and 1. In the former case, 0 constitutes the residue, in the latter, the form.

Take a positive Number N and its symmetric counterpart −N. To every Number situated 'between' Zero and N we can make correspond a Number situated 'between' −N and Zero: we just take its symmetric counterpart. In fact, it is clear that, where it is the case that Zero $< N_1 < N$, it is also the case that $-N < -N_1 <$ Zero. This can be verified by examining all possible cases of inequality between N_1 and N (see **13.13**), remembering that −N swaps the form and residue of N.

There are thus 'as many' Numbers between −N and Zero as there are between Zero and N, because the function $f(N_1) = -N_1$ is a biunivocal correspondence between the two 'slices' of Numbers. But take care! The correspondence is not between two *sets*. The interval between Zero and N is not a consistent totality any more than the entire domain of Numbers is. This can easily be proved: taking, for example, the Number (2,(0)), we know that *all* Numbers of the type (W,(0)), where W is any ordinal whatsoever larger than 2, are smaller than (2,(0)). It is the law that we discovered in **13.16**: if the form stays the same and the matter is increased, the Number gets smaller. Meanwhile, all Numbers (W,(0)) are positive, since 0 is in their form. So there are 'as many' of these positive Numbers – that is, those situated between Zero and (2,(0)) – as there are ordinals larger than 2. But we know for sure that 'all ordinals larger than 2' is an inconsistent multiplicity.

Keeping this in mind, we can allow ourselves to visualise symmetry in the following way, the axis being that of Numbers taken according to their order:

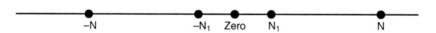

This justifies our speaking of a symmetry whose centre is Zero.

16.10. The ordinals

We announced a long time ago (see for example 8.8) that the ordinals, which constitute the stuff of the being of Numbers, can also themselves be represented as Numbers. What do the Numbers that represent ordinals look like?

Let's consider the Number (W,W), whose matter is the ordinal W and whose form retains *all* of this matter. In other words, this is a case of a maximal numerical section, or of exhibiting – as certain contemporary artists have done – the raw material alone as the 'work'. The most interesting thing is to compare the Number (W,W) with the Number (W,0), whose form is void. In both cases, we feel that the act is somehow null. But the two nullities are distinct. The Number (W,W) treats *the whole of the matter as a form*, whereas the Number (W,0) *does not inscribe any form in the matter*. The immediate result is that (W,W), for any W other than 0, is a positive Number, whereas (W,0) is a negative Number (remembering that 0 *is not* an element of 0, and that therefore 0 is not in the form of (W,0)). We discern a certain positivity in the first gesture which designates the matter as form, whereas the second, overwhelmed by the matter, is unable to designate anything whatsoever.

But if (W,W) is treated as a positive production, the assumption of a matter as form, it remains nevertheless a fact that this production repeats the ordinal–matter. This redoubling of the ordinal (as matter, then as form) legitimates our treating Numbers of the form (W,W) as the Numerical representatives of the ordinals.

We will therefore say the following: *An ordinal W is presented as Number in the form (W,W); that is, the Number whose matter is W and whose form is W*. This presentation is the ordinal 'itself', but *thought as Number*.

16.11. To be sure that this is the ordinal 'itself', we must explicitly prove that the order of Numbers respects the order of the ordinals, which is belonging. In other words, that if it is the case – ordinals being thought in their own domain – that $W_1 \in W_2$, then it is also the case – ordinals being thought as Numbers – that $(W_1,W_1) < (W_2,W_2)$.

This is obviously the case. Because the discriminant of (W_1,W_1) and (W_2,W_2) is necessarily the smallest ordinal to belong to W_1 and not to W_2, or to belong to W_2 and not to W_1. If $W_1 \in W_2$, this smallest ordinal is precisely W_1, which belongs to W_2 but cannot belong to itself. Now W_1 is outside the matter of (W_1,W_1), and it is in the form W_2 of (W_2,W_2). So it is indeed the case that $(W_1,W_1) < (W_2,W_2)$.

Thus the order of the ordinals thought as Numbers, in the formal redoubling of their material being, is the same as the order of ordinals thought in their being, as transitive sets all of whose elements are transitive. The Numerical representation of the ordinals is structurally isomorphic to the ordinals. This being so, there is no reason why we should not consider that the ordinals '*themselves*' are inscribed, identically represented, in the order of Numbers.

16.12. From the fact that an ordinal is a Number of the form (W,W), three consequences can be drawn:

1   Every sub-Number of an ordinal is an ordinal. For, if (W,W) is
    an ordinal, a sub-Number is of the form (w_1, w_1), where $w_1 \in$ W.
    It is therefore the ordinal w_1.
2   All these sub-Numbers will manifestly be ordinals smaller than
    the initial ordinal. It follows that they are *all* in the low set of
    the initial ordinal, and that the high set, generally composed of
    sub-Numbers larger than the Number, is empty here. This is a
    characteristic property of ordinals thought as Numbers. Gener-
    ally speaking, a sub-Number of the high set is a sub-Number
    N/w_1 such that w_1 is in the residue of N. But, in the case of
    an ordinal (and this could be a definition of the ordinals), *the
    residue is empty*. The high set of an ordinal is therefore also
    empty; and, conversely, if the high set of a Number is empty,
    then its residue is empty: its form coincides with its matter; it
    is an ordinal. The canonical presentation of an ordinal will
    therefore be of the form Lo(W)/0. But what is more, as the
    low set has for its elements *all* ordinals smaller than W, it is,
    as a set, identical to W (every ordinal is the set of all the
    ordinals smaller than it, **11.2**). Finally, the canonical representa-
    tion – most distinctive – of an ordinal W thought as Number
    is simply W/0.
3   The symmetric counterpart of an ordinal (W,W) is obtained by
    swapping the residue and the form. Now, the residue is empty.
    So it is the void that will be substituted for the 'total' form
    that is W: the symmetric counterpart of (W,W) is the Number
    (W,0). Thought as Number, an ordinal W allows of a symmetric
    counterpart, so we can freely speak of the Number –W.

It is clear that every ordinal apart from 0 is a positive Number, since
its form, W, contains 0 as an element. The symmetric counterpart
of every ordinal other than the void is therefore a negative Number,
as can be seen directly in writing (W,0). It will be found, moreover,
that all the properties of an ordinal W are inverted by the passage
to –W. So that now every sub-Number of –W is the symmetric
counterpart $-w_1$ of an ordinal w_1 smaller than W; and it is the low
set of –W that is void, since – the form of –W being void – every
sub-Number of –W is larger than it; and, finally, the high set of –W
is identical to –W, with the result that the canonical representation
is: 0/–W.

We are thereby assured that ordinals are Numbers.[1] But what is
more, grasped in terms of Numericality, the ordinals are symmetricis-
able: we have opened up on the other side of Zero (which is the
ordinal 0, thought as Number) an immense space where will be

inscribed those formerly unthinkable entities: natural multiplicities submitted to the negative. Numericality is capable of *symmetricising* nature.

16.13. Positive and negative whole numbers
The natural whole numbers, thought in their being, are none other than the finite ordinals, which is to say the elements of ω, the first limit ordinal. In fact we have already given their definition and discussed their operational dimensions in chapter 11.

This preceding work already settles the question, then: thought as Numbers, *natural whole numbers are of the type* (n,n), *where* n *is a finite ordinal.* Evidently, they are all positive. The order of natural whole numbers qua Numbers coincides with the order of natural whole numbers that we already know, the order according to which every schoolboy says that n is larger than p. For we know that, if $p \in n$ – which is the ontological version of traditional order – then $(p,p) < (n,n)$ in the order of Numbers. We therefore have the right to write the Number (W,W) as W, to indicate that an ordinal 'itself' is being inscribed in the domain of Numbers. We therefore write a natural whole number, thought as Number, as n.

The sub-Numbers of a natural whole Number are the finite ordinals smaller than it, therefore the natural whole numbers smaller than it. If n is this Number, these will be natural whole Numbers $(0,0),(1,1),\ldots,((n-1),(n-1))$, which we could also write as $0,1,\ldots,(n-1)$. Taken together, they form the low set of n. The high set of n is empty, and the canonical representation of a whole n, thought as Number, is $(0,1,\ldots,(n-1))/0$. Since n's elements are precisely $0,1,\ldots,(n-1)$, the low set whose elements they are can be written as $n/0$. (NB This is not circular, because, considered *as a set*, n does not contain itself as an element).

The symmetric counterpart of a natural whole number is a Number of the form $(n,0)$, where n is a finite ordinal. We write it $-n$, we say 'minus n'. We posit that *a Number is a whole negative Number if it is the symmetric counterpart of a natural whole Number, that is, one which takes the form* (n,0). The sub-Numbers of a negative whole number $-n$ are all the whole numbers $-p$, where $p \in n$. Taken together, they form the high set of $-n$, whose low set is empty. The canonical representation of a negative whole Number is therefore ultimately written as $0/-n$.

In order to confirm the complete identity of the traditional positive and negative whole numbers and of the positive and negative whole Numbers, it must obviously be the case that operations on these Numbers coincide, as order did, with operations on numbers. If for example we define an addition $N_1 + N_2$ on Numbers, then the result of this operation in the specific case of whole Numbers, hence in the case of an addition of the type $m + n$, should be 'the same Number' as the whole number which, in the calculations of our schooldays, corresponded to the addition of these two whole numbers. These operational verifications will be carried out in chapter 18.

So far as the inscription within the Numbers of natural whole numbers thought in their being is concerned, our task is complete.

16.14. Dyadic positive rational numbers

We have already spoken of rational numbers in relation to Dedekind cuts (compare **15.5**): a positive (or null) rational number is a fraction or relation $\frac{p}{q}$ of two natural whole numbers, which is to say a pair $\langle p,q \rangle$ of whole numbers. The first is called the *numerator*, the second the *denominator*. The numerator can be null (identical to the empty set), but it is prohibited for the denominator to be 0 (we know that the relation $\frac{p}{0}$ is 'undetermined').

We have no desire here to enter into a rigorous introduction to these traditional numbers (in fact, here we must consider fractions as irreducible, impossible to simplify). The intuitive idea of the fraction will suffice for us.

It is evident that the natural whole numbers are a subset of rational positive or null numbers; we just need to take a rational in the form $\frac{n}{1}$ to obtain n. In other words: a whole number is a rational of the type $\langle n,1 \rangle$.

The classical order of the rationals has the fundamental property of being a dense order. In other words (see **15.5**), given two rationals $\frac{p_1}{q_1}$ and $\frac{p_2}{q_2}$ such that $\frac{p_1}{q_1} < \frac{p_2}{q_2}$, however 'near' these two numbers might be, there always exists a third (and, from there, an infinity of them) which comes between the two initial numbers: there is a $\frac{p_3}{q_3}$ such that $\frac{p_1}{q_1} < \frac{p_3}{q_3} < \frac{p_2}{q_2}$.

A *dyadic rational number* is a number of the form $\frac{p}{2^n}$ whose denominator is a power of 2. Or, in our paired version, a rational number $\langle p,2^n \rangle$.

Dyadic rational numbers themselves form a dense subset of the rationals: if r_1 and r_2 are rationals such that $r_1 < r_2$, a dyadic rational can always be intercalated between them.

The important thing for us is that every sequence of augmenting rationals $r_1 < r_2 < \ldots < r_n < \ldots$ can be 'replaced' by a sequence of dyadic rationals $d_1 < \ldots < d_n < \ldots$: take the dyadic rationals situated 'between' r_1 and r_2, then r_2 and r_3, etc. We can also say that the dyadic rationals form a 'basis' for all the rationals. More specifically, a non-dyadic rational number can be 'approached' as closely as you like from a dyadic one, because you can always lodge a dyadic in-between r and $r + r'$, however small r' might be.

16.15. We have, then, the following statement, perhaps the most important in the process of the representation of (traditional) numbers as (ontological) Numbers:

> *Every dyadic rational number can be represented as a Number of finite matter, and every Number of finite matter represents a dyadic rational number.*

16.16. How, in general, is a Number of finite matter presented? In the form $(n,(p_1,p_2, \ldots p_i))$, where the whole numbers p_1, p_2, etc. which make up its form are whole numbers smaller than n, the matter of the Number. Since we are keeping to rational positive numbers, we will consider here only Numbers of positive finite matter, that is, Numbers which have 0 in their form.

The subtle idea that guides the 'projection' of these Numbers of finite matter into the dyadic positive or null rationals is the following. Let n be the matter of the Number. We take all the elements of this matter in order, from 0 to $n - 1$, which is the largest whole number contained in n. In so far as we stay in the location of the first element, 0 – which is the form, since the Number is positive – we attribute the value 1 to the whole number in question. Say that we come across the first element of n – say p – not to have the same location as 0, in other words the smallest whole number p in the matter of the Number to be in the residue. We attribute to this whole number the value $-\frac{1}{2}$. After this, we generally attribute to whole numbers q which follow the value $\frac{1}{2^{q-p+1}}$ if they are in the form, the value $-\frac{1}{2^{q-p+1}}$ if they are in the residue.

Finally, the value of the last term beyond p (p being still the first term which does not have the same location as 0, if it exists), the value attributed to $n - 1$, then, will be $\frac{1}{2^{n-p}}$, with or without the sign depending on whether $n - 1$ is in the residue or in the form.

Or, once again: a belonging to the residue will always be affected by the sign $-$. In traversing in order all the elements q of n, in so far as one remains within the form, which is the location of 0, each element is counted for 1, for a whole value. As soon as the location changes, we count the elements for a dyadic rational of the form $\frac{1}{2^{q-p+1}}$, where p is the first for which the location changes, from now on adding the sign $-$ whenever this location is the residue.

Finally, we associate with the initial Number of finite matter the rational number obtained from the sum (in the usual sense) of all the values thus attributed to the elements of n. This rational number is dyadic, since all the denominators in question are dyadic, and since – as every schoolchild knows – to add fractions, one takes as denominator the smallest common multiple of the denominators. Now, the smallest common multiple of powers of two is a power of two.

16.17. Let's give an example of the procedure. Take the Number $(5,(0,1,3))$, whose (finite) matter is the ordinal 5 and whose form contains 0, 1 and 3. The residue is thus composed of 2 and of 4.

Since 0 is in the form, we give it the value 1.

Since 1 is also in the form, we give it the value 1.

The location changes with 2, which is in the residue. We give it the value $-\frac{1}{2}$.

3 is in the form; we give it the value $\frac{1}{2^{3-2+1}} = \frac{1}{2^2}$.

4 is in the residue, we give it the value $-\frac{1}{2^{4-2+1}} = \frac{1}{2^3}$.

So, in the end, the rational number corresponding to the Number $(5,(0,1,3))$ will be obtained from the sum:

$$1 + 1 - \frac{1}{2} + \frac{1}{2^2} - \frac{1}{2^3} = \frac{13}{2^3}$$

We can see very well that this is indeed a dyadic rational.

16.18. In order better to exhibit the construction of this correspondence, which bears witness to an isomorphy, an identity of being, between positive Numbers of finite matter and positive dyadic rationals, we will formalise things a little. We will then see clearly that we are dealing with an inductive definition, a definition by recurrence.

Take a positive Number of finite matter. We will define by recurrence the following function f, defined on the elements of the matter n of the Number:

RULE 1: $f(0) = 1$.

RULE 2: $f(p + 1) = 1$, if $f(p) = 1$ for all whole numbers up to and including p, and if $p + 1$ is in the form of the Number.

RULE 3: $f(p + 1) = -\frac{1}{2}$ if all the whole numbers up to and including p are in the form and $p + 1$ is in the residue.

RULE 4: $f(p + 1) = \frac{1}{2^{q+1}}$ if the value of p is $\frac{1}{2^q}$ or $-\frac{1}{2^q}$ and $p + 1$ is in the form.

RULE 5: $f(p + 1) = -\frac{1}{2^{q+1}}$ if the value of p is $\frac{1}{2^q}$ or $-\frac{1}{2^q}$ and $p +$ 1 is in the residue.

These rules will allow us to calculate the rational value of f for all the elements of n, the matter of the initial Number. Using Ra(N) to denote the dyadic rational that corresponds to N, we then posit that:

$$Ra(N) = f(0) + f(1) + \ldots + f(n - 1)$$

The sign + indicates here the algebraic sum in the normal sense.

It is clear that Ra(N) is a dyadic rational.

16.19. Let's proceed with the calculation of another example, the Number $(4,(0,1,3))$, which is, of course, a positive Number of finite matter:

$f(0) = 1$ (by rule 1).

$f(1) = 1$ (by rule 1).

$f(2) = -\dfrac{1}{2}$ (by rule 3; 2 is in the residue).

$f(3) = \dfrac{1}{2^{1+1}} = \dfrac{1}{2^2}$ (rule 4; 3 is in the form).

So:

$$Ra((4,(0,1,3))) = f(0) + f(1) + f(2) + f(3).$$

$$Ra((4,(0,1,3))) = 1 + 1 - \frac{1}{2} + \frac{1}{2^2}.$$

$Ra((4,(0,1,3))) = \dfrac{7}{2^2}$, which is a dyadic rational, as we said it would be.

16.20. Whole ordinal part of a Number

It might appear strange peremptorily to change the procedure when we get to the first whole p that doesn't have the same location as 0 in the Number of finite matter under consideration. Gonshor realises this: 'The whole idea of a shift from ordinary counting to a binary decimal computation at the first change in sign may seem unnatural at first. However, such phenomena seem inevitable in a sufficiently rich system.'[2] This explanation of Gonshor's – more of an apology, really – is a little quick.

To find the true underlying concept, we should ask what is actually represented by the p first consecutive ordinals of a Number N of finite matter which have the same location as the initial term 0. Assume once more the positive case (0 located in the form). If we partition N at point p (the first ordinal, in the order of ordinals, to change location), we obtain the sub-Number N/p *all of whose elements have the same location as 0*. It is clear that, since this location is the form, N/p is the whole positive Number p, that is, the Number whose matter is p and whose form is made up of the elements of p. The function f will attribute the value 1 to all these elements, and the sum of the values $1 + 1 + \dots$ will give the 'classic' whole number p. Which, we can add immediately, is an algebraic sum of dyadic rationals of the type $\frac{1}{2^q}$ or $-\frac{1}{2^q}$, where q is no more than 1. It follows that Ra(N) will be the sum of the whole p and a negative dyadic fraction between −1 and 0 (at least, unless it happens to be a whole number). Finally, p is a type of *whole part* of the positive rational Ra(N), that is, the natural whole number closest to Ra(N) 'from above': $(p - 1) < \text{Ra(N)} < p$.

From the point of view of Number, in fact, p is the largest sub-Number of N *to be an ordinal*, since 'being an ordinal' means precisely being a Number all of whose matter is in its form. That the location changes at point p (p is in the residue) means precisely that N/p + 1 is no longer an ordinal either, since p, an element of its matter, is in the residue. It is therefore even more fitting to say that p is a 'whole part' of N. By which we mean: the largest whole number p belonging to the matter of N and such that the sub-Number N/p is the ordinal p. Or even more simply: the whole part of N is the largest ordinal to be a sub-Number of N.

Now the procedure becomes clearer: it works *firstly* by making correspond, via f, the elements of the whole part of N and the whole part 'from above' of the dyadic rational Ra(N). The 1 values are used to do this. And *then* it is a question of calculating the remainder, which is less than 0, but more than −1, and to do this we use dyadic fractions of the type $\frac{1}{2^q}$ or $-\frac{1}{2^q}$, q indicating the rank of the ordinal in question *beyond the whole part* p. There is no 'unnatural' mystery in all of this, but rather a profound logic.

16.21. We can generalise these remarks. Given a positive Number N of matter W, we will call *whole ordinal part* of N the largest ordinal $w_1 \in W$ such that the sub-Number N/w_1 is the ordinal w_1.

The attentive reader will balk at this: how can we speak so freely of the 'largest ordinal' to satisfy a property? Doesn't the existence of limit ordinals militate against any such claim? Where ordinals are concerned, only minimality is at work.

The remark is well taken. We will have to reformulate our definition, then, and posit the following: *The whole ordinal part of a positive Number is the smallest ordinal located in the residue.* Since the Number is positive, 0 is located in the form. The smallest ordinal located in the residue is thus indeed the first ordinal, in the ascending order of the ordinals of the matter of N, all of whose elements are in the form, although it itself is in the residue. These elements constitute the whole ordinal part of N. Here is a case where 'the largest' translates as 'the smallest'.

If w_1 is the whole ordinal part of a positive Number N then, just as in the above case, $N < w_1$, since the sub-Number w_1 is, considered as an element of the matter of N, in the residue of N, whereas it is outside its own matter.

It can also be said that the whole ordinal part of a positive Number is in the high set of that Number.

If w_1 is a successor ordinal, once again we find the 'framing'[3] of the endpoint. Let w_2 be the predecessor of w_1; this gives $w_1 = S(w_2)$. Since w_1 is the *smallest* ordinal to be in the residue, its predecessor w_2 must be in the form. Of course, since all the elements of w_2 are elements of w_1 (transitivity of ordinals), and all the elements of w_1 are in the form, all the elements of w_2 are too; so N/w_2 is the ordinal w_2. And, given that this ordinal is outside its own matter and in the form of N, then $w_2 < N$, and so finally $w_2 < N < S(w_2) = w_1$. This is the interval we are looking for.

If, on the other hand, w_1 is a limit ordinal, it will certainly always be the case that $N < w_1$, but we would search in vain for the largest ordinal smaller than N, because on the other side of w_1 there is no 'predecessor'. N would then have a singular position: smaller than a limit ordinal, *it would be larger than all the ordinals smaller than this limit ordinal.* It would come to insert itself in that space we thought was 'filled in' by the ordinals that precede the limit, the space 'between' a limit ordinal and the infinity of successor ordinals of which it is the limit.

16.22. Let's give an example. Take the Number $N = (S(\omega), S(\omega) - (\omega))$, whose matter is the successor of ω and whose form is all of that matter except for ω itself, which is the only element of the residue. The limit ordinal ω, being the first ordinal in the matter of N to be in its residue, is the whole ordinal part of N. It is indeed the case that

$N < \omega$, since their discriminant is ω, which is in the residue of N and outside the matter of ω. What is more, for every element of ω – that is, for every natural whole number n – it is the case that $n < N$, since n is outside the matter of n and in the form of N. The Number N is thus at once smaller than the first limit ordinal ω and larger than all the natural whole numbers n of which ω is the limit! This shows to what extent the domain of Numbers saturates that of the ordinals, which it contains: there are 'many more' Numbers than there are ordinals.

We can also say that N is 'infinitely near' to ω, far nearer than even the most immense of the whole numbers could be. This notion of 'infinite proximity' is of a prodigious philosophical interest. It opens up new spaces for exploration in the endless kingdom of Number. We shall undertake these explorations a little later.

16.23. Sequence and end of the dyadic rationals

We have at our disposal a function Ra(N) which makes a dyadic rational correspond to every Number of finite matter. The whole numbers are included in this correspondence, because the positive whole number n thought of as Number will correspond, through the function Ra, to the sum $1 + 1 + \ldots + 1$ n times – that is exactly the Number n, since, if a Number is a natural whole number, then *all* of its sub-Numbers are in its form. It would be better to say that the function Ra associates a dyadic rational with every Number of finite matter – even if this Number is whole.

To complete the work, and to conclude that the dyadic rationals 'themselves' are represented in Numbers, we must:

- confirm that the order of Numbers of finite matter is isomorphic with the customary order of corresponding dyadic rationals, so that, if $N_1 < N_2$ in the order of Numbers, then $Ra(N_1) < Ra(N_2)$ in the normal order of rationals; this amusing mathematical exercise is sketched nicely in the note;[4]
- prove that *all* the dyadic rationals are obtained through the function Ra applied to Numbers of finite matter; this comes down to proving that every positive dyadic rational can be put in the form of the algebraic sum of a certain whole number (its whole part 'from above') and dyadic rationals of the form $\frac{1}{2^q}$ or $-\frac{1}{2^q}$; because, once this is done, one can reassemble the Number N, whose value for Ra is the rational thus dismembered;[5]
- prove that the operational dimensions of the rationals – addition, multiplication, division, in brief, everything that gives them the algebraic structure of a field, are isomorphic to the same

operations defined for Numbers and applied to Numbers of finite matter; this relates to the examinations made in chapter 18, with one obvious exception: in order to have *negative* dyadic rationals, the procedure of symmetricisation would be used, which defines the general manner of passage to the negative: inversion of swapping form and residue. Of course, we will still be dealing with a Number of finite matter (but this time with 0 in the residue).

As far as the ontological side of things is concerned, we have attained our goal. A dyadic rational, thought in its being, inscribed as Number, has a very simple intrinsic definition: its matter is finite.

As far as being is concerned, that, this clarifies however dense the rationals might be, even to the point of an infinite swarming between two consecutive whole numbers, they nevertheless belong to the finite. The numerical ontology of the infinite begins with real Numbers.

16.24. Real numbers
We know that real numbers provide the model for the geometrical 'continuum': their figure is that of the points of a line. It is the real numbers that have subtended the entire edifice of analysis, *chef-d'oeuvre* and keystone of modern mathematical thought, since Newton and Leibniz.

For a long time, the continuum and the functions corresponding to it were thought either in terms of geometrical constructions (Greek and pre-classical age), or in a primitive and pragmatic fashion (eighteenth and nineteenth centuries). The emergence of a rigorous concept of reals as entities with which one can calculate took place slowly during the course of the nineteenth century, beginning with Cauchy, and with Dedekind representing a decisive step.

Because it is the closest to that which governs the definition of reals in the field of Numbers, we will recall briefly the construction of real numbers by means of 'cuts', as invented by Dedekind.

16.25. We will begin with dyadic rationals, which we can use here in place of rationals as such, in view of the remark made in **16.14**. Take two sets of dyadic rationals B and A such that every rational in B is smaller than every rational in A. We can say both that B has no internal maximum (for every dyadic rational r_1 in the set there is an r_2 in the set such that $r_1 < r_2$); and that A has no internal minimum. Suppose now that the following relation holds between B and A: there always exists a dyadic in B that is 'as close' as one likes to a dyadic

of A. In other words, if r_2 is a dyadic in A and r a dyadic *as small as one likes*, there will always exist a r_1 in B such that the difference between r_2 and r_1 is less than r.

The situation can be visualised as below, by representing the dyadic rationals as points on a line:

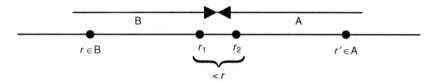

We can see clearly that B 'rises' without ever entering into A, that A 'descends' without entering into B, and that therefore the two sets are as close to each other as can be, without ever 'touching'.

Thus did Dedekind *define* a real number as *the* point situated exactly 'between' B and A; that is, the element, *created* in this process, which is simultaneously larger than any element of B and smaller than any element of A. We can identify this element as the point of the *cut* of B and A.

It is characteristic of this method that it treats the cut not as a state of things in a pre-given universe (which is how we treated it for Numbers, see **15.6**), but as a procedure, defining a mathematical entity that does not pre-exist this procedure. To begin with, there are only rationals. And, if the cut is not a rational (it could be, if the upper limit of B and the lower limit of A coincided), then it constitutes in itself the name, or form of presentation, of a 'being' which inexists in the field of rationals. Therefore the reals are operational productions here; they *sign*, coming forth from non-being, the fictive point where B and A are touched by the interposition between them of this fiction. Into that place, where there was nothing but the miniscule void that separates two sets as close as can be, comes the real, which stops up this void by *realising* a cut as number.

16.26. Fictions have no place in the ontological conception of Number. If the classic real numbers, those which realise cuts in the dyadic rationals, inscribe themselves in the domain of Numbers, it is because they exist and are distinguished by some property. They cannot irrupt from inexistence, in the form of mere *names* of a lacuna. According to an ontological conception of Numbers, every Number *is*, none results or is resolved in the name of an operation. We do battle here against a dominant nominalism, and we do so in the field of number, so commonly taken for an operational fiction.

16.27. In fact, our definition of real numbers as Numbers is quite limpid:

A Number is a real number if it is either of finite matter or of matter ω, *and if its form and its residue are infinite.*

In what follows we will substantiate this definition, which represents real numbers 'themselves' in the domain of Numbers.

16.28. The 'projection' of this definition into the concept of reals as cuts is basically very simple.

If a Number is of finite matter, then it is, as we have seen, a dyadic rational.

If a Number is of matter ω, then all of its sub-Numbers are of lesser matter than ω, and therefore of finite matter, since ω is the smallest infinite ordinal. So all of its sub-Numbers are dyadic rationals. More specifically, its low set and its high set are sets of dyadic rationals. And, since every Number is the cut of its low set and its high set, a Number of matter ω can be represented as the cut of two sets of dyadic rationals. Or, once again, a real number thought as Number is a Number whose canonical presentation Lo(N)/Hi(N) is made solely from dyadic rationals.

Finally, if a Number of matter ω has an infinite form and an infinite residue, we avoid its low set and high set having *internal* maxima. Because, if the form of N is finite, since it is composed of whole numbers (the matter being ω), it admits of a largest element, say the whole number p. The cut of N at point p defines the sub-Number N/p, which is obviously the largest sub-Number whose discriminant with N is in the form of N, and therefore the largest sub-Number in the low set of N. And, if the residue is finite, there exists a number p such that N/p is the smallest element of the high set of N. *A contrario*, if both the form and residue of N are infinite – are sequences of whole numbers without internal maxima – then the low set does not have a maximum term, nor the high set a minimum term.

We thus find ourselves precisely in the conditions of the Dedekind cut: disjoint ascending and descending sets of dyadic rationals with no maximum or minimum. Except that *what we characterise as 'reals' are particular, already existing Numbers*, whereas Dedekind installs them as a fiction at the void point of a cut. For us, a real will be that unique Number *of minimum matter* situated exactly between two sets of dyadic rationals which can be shown to be its low set and its high set, and therefore to be sets of sub-Numbers.

It is particularly reassuring to remark that, in the definition of reals as Numbers, everything remains immanent. Dedekind cuts designate the fiction of a number external to two sequences of rationals, as the point of contact of these sequences. Whereas, on the other hand, the sets of dyadic rationals that we use are composed *of sub-Numbers of a real Number*. This immanentisation of procedures is typical of the ontological approach, that approach which captures the being of Number. To see if a Number that is not a dyadic rational is a real number, it suffices to examine it according to its three components:

- its matter must be ω;
- its form must be infinite;
- its residue must be infinite.

This alone allows us to conclude. *Then* we can state that the Number is the cut of two sets of dyadic rationals, and that therefore it is indeed a real number (in the classic sense). But, all the same, we have remained within Number, since dyadic rationals are sub-Numbers of a Number.

The immanence of the thinking of being has not faltered for a moment in this approach to the traditional real numbers grasped in the space of Numbers. The characterisation of a type of pure multiple has been substituted for operational fictions. And real numbers are no more mysterious here than whole numbers or rationals. Their sole peculiarity is that they mark the moment where our passage through Numbers prompts us to envisage infinite matters. From this point of view, the ontological singularity of the reals in relation to the wholes and the rationals can be summed up in one word: infinity. This alone clarifies, irrespective of all complexities of construction, with an eye only to that in which the numerical section operates, the fact that real Numbers are exemplarily *modern*.

16.29. We now find ourselves in possession of a concept of Number that subsumes as particular species the natural whole numbers, the whole positives and negatives, the rationals, the reals, and the ordinals.[6] We have overcome the modern resistance to a unification of the concept of Number (see **1.8**). But, in the process, we have already seen that this concept also subsumes other Numbers, that the historical deduction from the domain of Numericality is very much limited. Rationals and reals cover the totality of Numbers of finite matter and only some Numbers of matter ω. It is as if *our* thinking has so far only brought to light a minute initial segment of that which being proffers in terms of possible numerical access to pure multiplicities. The future of the thinking of Number is limitless.

16.30. Infinitesimals

We remarked in **16.22** that it would be possible to find a Number N at once smaller than ω and larger than all the finite ordinals whose limit is ω. This Number could perhaps be said to be 'infinitely close' to ω, and it puts us on the way to a concept of infinitesimal Number.

The idea of infinitely small number, freely employed by seventeenth- and eighteenth-century mathematicians, was dismissed in the nineteenth century for its obvious inconsistencies. It was replaced by the concepts of the limit (Cauchy) and of the cut (Dedekind). It reappeared around thirty-five years ago, in the singularly artificial, but consistent, context of the pure logic of models: Robinson's nonstandard analysis.[7] In the domain of Numbers, 'infinitely small numbers' or infinitesimals abound in the most natural fashion. It is by means of them that we will complete this diminutive journey through the enchanted kingdom of Numbers.

16.31. Consider the Number $i = (\omega,(0))$, whose matter is ω and whose form, the singleton of the void, has the void as its only element. It is a positive Number, since 0 is in its form.

Now this positive Number, even if its matter is the same as that of real Numbers, *is smaller than every positive real Number*.

In fact, if a real Number is positive, 0 is in its form, as is the case for i: 0 does not discriminate between i and a positive real Number. All the whole Numbers other than 0 being in the residue of i, the discriminant of i and a real Number R will be the first whole Number apart from 0 to figure in the form of R. Such a Number necessarily exists, since the definition of the reals dictates that the form of R should be infinite. And, since this discriminant is in the residue of i, i is smaller than R. Therefore there exists a Number i such that $0 < i < R$ for every real Number. This i is situated 'between' Zero and all real numbers thought as Numbers. We will say that it is infinitesimal for the reals.

16.32. Generalising this definition: We say that a *set* of positive Numbers, all of the same matter, tends rationally towards Zero if, for every dyadic positive rational r, as close to Zero as you like, there exists a Number N_1 of the set situated between Zero and r. In other words: for every dyadic rational r, there exists N_1 belonging to the set such that $0 < N_1 < r$. Note that the classic notion of 'tending towards' is here relativised to dyadic rationals. In the limitless domain of Numbers, we must indicate which scale of measurement is being employed, because, as we will see, it is *always* possible to find a still finer scale.

It is obvious that the set of real positive Numbers tends rationally towards Zero. Other sets of Numbers can be found which tend rationally towards Zero, for example positive Numbers of the type $(S(\omega),(0, \ldots))$, whose matter is $S(\omega)$ and whose form contains at least 0.

We can say, then, that:

A Number is infinitesimal for a set of Numbers that tends rationally towards Zero if it is:

– *of the same matter as the Numbers of the set;*
– *positive;*
– *smaller than all the Numbers in the set.*

So it is that the Number $(\omega,(0))$ is infinitesimal for the set of real Numbers. On the other hand, there is no infinitesimal for the set of Numbers $(S(\omega),(0, \ldots))$, precisely because this set *contains* the very Number $(S(\omega),(0))$ that is the smallest positive Number of matter $S(\omega)$.

The limiting of the concept of infinitesimal to Numbers of the same matter as the Numbers of the set that tends towards Zero is necessary because, if this restriction were not in place, *there would be as many infinitesimals as we wished*. It would suffice to augment the matter: the Number $(S(\omega),(0))$ is positive, and it is certainly smaller than every positive Number whose matter is ω. In particular, it is smaller than the infinitesimal $i = (\omega,(0))$, because the discriminant is ω, which is outside the matter of i and in the residue of $(S(\omega),(0))$. We see to what extent our concept of the infinitesimal is relative: the density of order over Numbers means that, however 'relatively' small a positive Number might be, there still exists an inconsistent multiplicity of Numbers situated between it and Zero.

We can, if we wish, retain the classic definition: every positive Number smaller than every positive real is infinitesimal. But then we will see the infinitesimals grow and swarm uncontrollably. The 'shores' of Zero contain 'as many' Numbers as the entire domain of Numbers. Because, at the point where multiple–being as such inconsists, the notion of 'as many' loses all meaning.

16.33. Cuts of cuts

Take the Number $C = (\omega,(0,1))$, whose matter is ω and whose form is limited to the wholes 0 and 1. This Number is not real, since its form is finite. It is positive, since 0 is in its form. How can it be situated amongst the reals, to which its matter belongs?

A positive real which does not have 1 in its form is certainly smaller than C: the discriminant is 1, which is in the residue of such a real and in the form of C.

A positive real which does have 1 in its form is certainly larger than C. For all whole numbers larger than 1 are in the residue of C, whereas some of them are certainly in the form of a real, since this form is infinite. The discriminant will be the smallest whole larger than 1 to be in the form of the real, and, since it is in C's residue, C will be smaller.

C therefore is situated precisely between the reals which have 1 in their residue and the reals which have 1 in their form. Now these two classes operate a partition into two of the positive reals, a partition which is ordered (all the positive reals which have 1 in their residue are smaller than all the positive reals which have 1 in their form). We can, then, perfectly lodge a Number 'between' two disjoint classes of reals, in the caesura of a partition of reals. And, since the reals are themselves cuts of rationals, the Number C will be a cut of cuts.

Generally speaking, given an organised partition into two of a set of Numbers 'of the same type', that is to say, defined by cuts or canonical presentations having this or that property (as we saw in defining the reals), we will call a 'cut of cuts' a Number of minimal matter situated in the caesura of the partition, being larger than all those in the lower segment and smaller than all those in the higher segment. The Number $(\omega,(0,1))$ is a cut of cuts in the numerical type 'positive real Numbers'.

The existence of cuts of cuts attests once more to the infinite capacity of Numbers – as coalescent as they might seem – for cutting *at a point* the ultra-dense fabric of their consecution.

16.34. So many other Numbers to visit and to describe! But works that take delight in this are beginning to appear. And the philosopher is not defined by curiosity; the journey is not a disinterested one. The philosopher must, before leaving the kingdom convinced that every number thought of in its being is a Number, descend back down to calculation. Or, rather, to the existence of calculation, because the philosopher is not a calculator either. But these numbers, from which our soul is knitted, the philosopher wishes to render over entirely, even as regards the derivation of their operational mechanism, to the immemorial and effectless transparency of Being.

4

Operational Dimensions

17

Natural Interlude

17.1. The domain of ordinals (and of cardinals) holds an extreme *charm* for thought. A proof by affect – by affection, even – of what I claim here, is that, on reflection, this charm is that of Nature itself: an abundant diversity and, at the same time, a mute monotony. Nothing is the same, everything goes to infinity, but one hears a fundamental note, a *basso ostinato*, signalling that these myriads of multiplicities and forms, these complicated melodies, proliferate the repose of the identical. If poets' metaphors take as their reference the sky and the tree, the flower and the sea, the pond and the bird, this is because they would *speak* this presence of the Same that the unlimited appearances of nature veil and reveal. In the same way, the ordinals, still singular in the infinity of their infinite number, in the inconsistency of their All, also repeat the transitive stability and the internal homogeneity of natural multiples, those multiples that they allow to be thought in their pure being. It is hard to tear oneself away from the intellectual beatitude brought on by the contemplation of the ordinals, one by one and as a 'set'. I think of the great Indian mathematician Ramanujan,[1] who held each whole number to be a personal friend. He was invested by this poem of Number, of which the Poem of nature is the symmetrical counterpart within language. He did not like to construct proofs, but rather, as a dreamer of the ordinal site, to *draw* in it with curves of recognition, which his colleagues regarded with some surprise. Coming from afar, in all senses of the word, he was not accustomed to our severe modern distinctions. He saw numbers directly for what they are: natural treasures, where being lavishes its multiple resource and its fastidious identity in the same gesture in which, for the poet, it arranges the 'correspondences' of sensibility.

17.2. We have at our disposal a concept of Number, and we know that this concept subsumes our traditional numbers. Wholes, rationals, reals, ordinals, thought in their multiple–being, are Numbers. It must now be shown – a slightly less rewarding task – that this concept subsumes our traditional numbers not only in their being, but also in their operations. As far as we may be from that sensibility that is ruled by counting, it must nevertheless be shown that it is possible to count with Numbers, and that this counting coincides, for the classical types of Numbers, with ordinary counting. We must cover algebra, addition, multiplication, etc. If we did not, then who would believe us when, speaking from the sole point of view of being, we said that these Numbers are numbers?

17.3. What is meant by 'operation', or calculation, is the consideration of 'objects' upon which one no longer operates one by one, but at least two by two: the sum of x and y, the division of x by y, etc. And, as the matter of Number is made of ordinals, it is to be expected that we have to deal with, to think, pairs of ordinals. So we will be happily detained for a few more moments in the enchanted domain of natural multiples. This whole interlude is dedicated to some reflections and propositions about pairs of ordinals, ordinals taken two by two. And, as we shall see, these couples are also totally natural: we can connect them back to 'single' ordinals via a procedure which in itself holds a great charm.

17.4. We will speak of *ordered pairs* of ordinals, which we denote by $\langle W_1, W_2 \rangle$. 'Ordered' meaning that one takes into consideration the order of the terms in the couple – we will thus speak of the first term, W_1, and the second, W_2 – which wasn't the case in our concept of the simple pair, denoted by (e_1, e_2) (compare 7.7), which was a pure 'gathering together' of two terms regardless of their order. Or, in other words: if W_1 and W_2 are different, then the ordered pair $\langle W_1, W_2 \rangle$ is not the same thing as the ordered pair $\langle W_2, W_1 \rangle$. In order better to distinguish the simple pair from the ordered pair, we will call the latter a *couple*.

We can also allow 'couples' of the type $\langle W_1, W_1 \rangle$. In such cases, W_1 occupies both the first and the second place.

The concept of ordered pair, or couple, plays a decisive role in mathematics: it underlies all thinking of relations and of functions.[2]

It can be reduced to a figure of the pure multiple, testifying to the fact that relations and functions do not depend on any sort of *additional* being apart from the multiple, that there is no ontological distinction between bound objects and the bond which binds them. But we will employ the concept here in its naïve sense.

17.5. We will call *maximal ordinal of a couple* $\langle W_1, W_2 \rangle$, and denote by Max $(\langle W_1, W_2 \rangle)$, either the larger of the two ordinals W_1 and W_2, if they are different, or, if the couple is of the type $\langle W_1, W_1 \rangle$ the single ordinal W_1 that figures in it. You are reminded (see **8.10**) that ordinals are totally ordered by belonging: if W_1 and W_2 are different, then one is necessarily smaller than the other (belongs to the other).

This most elementary notion of the maximal term of a couple will play a crucial role in what follows. It is important to get a firm grasp of it.

17.6. Take a couple of ordinals $\langle W_1, W_2 \rangle$, which we will denote by C_1, and another couple $\langle W_3, W_4 \rangle$, which we will denote by C_2. We will define an order-relation between these couples in the following way. We say that C_1 is smaller than C_2 and write $C_1 < C_2$, if one of the three following conditions is satisfied:

1   The maximal ordinal of the couple C_2 is equal to the maximal ordinal of the couple C_1. In other words: if $Max(C_1) \in Max(C_2)$, it is always the case that $C_1 < C_2$.
2   The maximal ordinal of couple C_1 is equal to the maximal ordinal of the couple C_2, but the first term of the couple C_1 is smaller than the first term of the couple C_2. In other words, in a case where $Max(C_1) = Max(C_2)$, if $W_1 \in W_3$, then $C_1 < C_2$.
3   The maximal ordinal of the couple C_1 is equal to the maximal ordinal of the couple C_2, and the first term of the couple C_1 is equal to the first term of the couple C_2, but the second term of C_1 is smaller than the second term of C_2. In other words, $Max(C_1) = Max(C_2)$ and $W_1 = W_3$, but $W_2 \in W_4$. In this case, $C_1 < C_2$.

Evidently, if none of these three conditions are satisfied, then the couples C_1 and C_2 must be identical: they have the same first term and the same second term. *A contrario*, if two couples of ordinals are different, either one is smaller than the other, or the other is smaller than it: the relation is *total*.

This order follows directly from employment of the operator $Max(C)$, or, if this yields only an identity, from the comparative

186 OPERATIONAL DIMENSIONS

examination, in this order, *firstly* of the first ordinal of each couple, *then*, if this examination too yields only an identity, of the second ordinal of each couple. The maximum trumps the first term, and the first term the second. Some examples:

- $\langle 6,0 \rangle$ is smaller than $\langle 0,7 \rangle$, because its maximum is 6, which is smaller than the maximum of the latter, which is 7;
- $\langle 0,\omega \rangle$ is smaller than $\langle 1,\omega \rangle$, because, their maxima being identical (it is ω), the first term of the former, 0, is smaller than the first term of the latter, 1;
- $\langle S(\omega),\omega \rangle$ is smaller than $\langle S(\omega),S(\omega) \rangle$, because, their maxima being equal (they are both $S(\omega)$, the successor of ω) and their first term likewise (it is $S(\omega)$ in both cases), the second term of the first couple, which is ω, is smaller than the second term of the other, which is $S(\omega)$.

Note that the couple $\langle 0,0 \rangle$ is the smallest couple of all, since its maximum, its first term and its second term are all equal to 0, which is itself the smallest ordinal.

It is also clear that couples form an inconsistent multiple, since, already, the ordinals themselves cannot form a set. In speaking of 'the' couples, but also of 'the' ordinals, or 'the' Numbers, we must always remember that we cannot attribute any property to whatever this 'the' designates: there is no question of a thinkable, or presentable, totality. In particular, if there exists a minimal couple for the order that we are going to define (it is the pair $\langle 0,0 \rangle$), there certainly is not a maximum couple for this order.

17.7. I leave the reader the task of showing that the relation between couples that we have just defined is a genuine order-relation (and therefore, essentially, transitive: if $C_1 < C_2$ and $C_2 < C_3$, then $C_1 < C_3$).

Far more interesting is the fact that it is a *well-ordered* relation. I have given the definition of this in 6.4: given any set whatsoever of terms well-ordered by a relation $<$, there exists one (and one only) element of that set that is *minimal* for the order-relation, which is *the* smallest element of that set.

Take any (non-empty) set E of couples of ordinals – that is, a set all of whose elements are couples of ordinals. Consider all those couple elements of E *whose maximum is minimal for E*. In other words all the couples $C \in E$ such that Max(C) is the smallest ordinal to figure in the elements of E as maximum of a couple. This is possible by virtue of the principle of minimality that characterises the ordinals (see **8.10**). Given the property 'being a maximal ordinal in a couple C which belongs to E', there exists a smallest ordinal to satisfy this property. We thus obtain a subset E′ of E, all of whose elements C have the same minimal maximum. Note that, because of the first of the conditions defining the order of couples, *all* the elements of E′ are smaller than *all* the elements that remain, that are in E–E′ (if any).

Now consider, in E′, the set of couples whose first term is minimal for E′. In other words, all the couples $C = \langle W_1, W_2 \rangle$ such that W_1 is the smallest to be found in all the first terms of the couples in E′. This is possible for the same reason as before: it suffices to consider the property 'being an ordinal that figures in a couple in E′ as the first term', and to take the minimal ordinal for this property. We will thus obtain a set E″ of couples having the same maximum (because they are in E′) and the same first term (minimal for E′). Note, in considering the second of the conditions defining the order of couples, that *all* the elements of E″ are smaller than *all* the elements that remain, which are in E′–E″, which themselves are all smaller than the elements of E–E′.

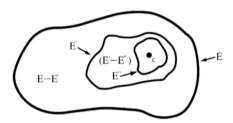

There is a sort of concentric embedding, where the couples of each inner circle are smaller than those of the exterior boundary.

Consider, finally, in E″, the property 'being an ordinal that figures in second position in one of the couples in E″'. There is a minimal ordinal for this property. But, this time, the set obtained consists of *one couple only*. This is because, in E″, the first term of the couples is fixed (it is the minimal first term for the couples in E′). In fixing the second term (as minimal for this place), one couple is entirely determined. But the others are themselves smaller than all the couples in E′–E″, which are smaller than the couples in E–E′. So the minimal couple obtained in E″ is in fact minimal in E–QED.

This property of minimality for the order of couples of ordinals grants us three essential freedoms:

1   Given a couple C, it is possible to designate the unique couple that will come directly after it in the order we have defined. To do so, it suffices to consider, in a suitable set that contains C, the subset of those that are larger than it. This subset will have a minimal element, which is the smallest one to be larger than C, and is thus the 'successor' of C.

2   If a property of couples defines a set (the set of couples which possess that property), then we can safely speak of the smallest couple in that set, and therefore of the smallest couple to possess that property.

3   Given a set of couples, we can speak of the upper bound of that set, as we can for sets of ordinals (see **12.16**): it suffices to consider the smallest couple that is larger than all the couples in the set.

'Well-orderedness' allows thought to move between *interior* minimality and *exterior* maximality: the smallest of a given set, and the first (outside) to be larger than all those in that set. The trap is to imagine that one thereby gains access to internal maximality: this is not at all the case because, for couples as for ordinals, that which goes to the limit is not internally maximisable.

17.8. We are speaking of succession and limit. Here we return, let us remark, to the disputations of chapter 9. Discovering the kinship between ordinals and couples of ordinals was our latent motive.

17.9. Let's begin with an example. What can we say of a couple of the form $\langle W_1, S(W_2) \rangle$, where W_1 is any ordinal whatsoever apart from 0, and where $S(W_2)$ is the successor of any ordinal W_2? Everything depends on the maximal ordinal in the couple. Suppose that W_1 is maximal and thus that $S(W_2) \in W_1$. If we compare the couple to all others that also have W_1 as their maximal ordinal, we see that it is:

–   larger than all those where W_1 only comes in second position in the couple (primacy of first position, condition 2 of the ordering of couples);

–   larger than all those which, in second position, have an ordinal smaller than $S(W_2)$ (third condition of order); in particular, it is larger than the couple $\langle W_1, W_2 \rangle$;

–   smaller than all those which have, in second position, an ordinal larger than $S(W_2)$. In particular, it is smaller than the couple $\langle W_1, S(S(W_2)) \rangle$, supposing that $S(S(W_2))$ remains smaller than W_1, thus leaving W_1's maximal status intact. But let's assume this hypothesis.

It seems clear from this that the couple $\langle W_1, S(W_2) \rangle$, given the assumed hypothesis, intercalates itself exactly between the couple $\langle W_1, W_2 \rangle$ and the couple $\langle W_1, S(S(W_2)) \rangle$. More specifically, we can say that it *succeeds* the first of the two couples.

If, on the other hand, we take the couple $\langle W_1, L \rangle$ where L is a limit, and still suppose that W_1 is maximal in the couple, we cannot determine a couple that it succeeds. This couple is certainly larger that *all* the couples of the form $\langle W_1, W_2 \rangle$ where W_2 is smaller than L (third condition of order). But L, a limit ordinal, precisely does not succeed any of the W_2 in question. There is therefore only one possibility: the couple $\langle W_1, L \rangle$ is the *upper bound* (see **N6**) of the set of couples $\langle W_1, W_2 \rangle$, where $W_2 \in L$, with W_1, of course, being maximal – that is to say, larger than L. We can also say that the couple $\langle W_1, L \rangle$ is the limit of the couples $\langle W_1, W_2 \rangle$ for W_2 less than L.

Finally, take the couple $\langle 0, S(W_2) \rangle$. The Max. of this couple is evidently $S(W_2)$. But it is certainly the *smallest* couple to have this Max. In fact, its first term is minimal (it is 0), so every couple C where $Max(C) = S(W_2)$ and where the first term is not 0 – therefore every couple of this sort other than our example – is greater than it.

Being the smallest couple whose Max. is $S(W_2)$, our couple must succeed the 'largest couple' – if it exists – whose Max. is immediately inferior. Note that these notions of 'larger' and 'immediately inferior' can be disrupted by the intervention of limit ordinals. All the same, this is not the case in our example: since the Max. of our couple is $S(W_2)$, an immediately inferior Max. exists: it is W_2. What would be the largest couple whose Max. is W_2? Obviously that couple whose first term is maximal (condition 2 of order). But the first term *of a couple whose Max. is* W_2 attains its maximum when it is equal to W_2. For, if it surpasses W_2, the Max. changes. So the couple that immediately precedes $\langle 0, S(W_2) \rangle$ in the order of couples is $\langle W_2, W_2 \rangle$. We can also say that $\langle 0, S(W_2) \rangle$ is the successor couple of $\langle W_2, W_2 \rangle$.

17.10. What we really want is to 'ontologise' couples of ordinals, as we did for ordinals: find an intrinsic characterisation, not bound to order alone, of successor couples and limit couples. The examples in the previous paragraph will guide us.

17.11. Let's begin with couples containing 0.

We have remarked that couples of the form $\langle 0, W_1 \rangle$, for all W_1 other than 0, are the smallest ones whose Max. is W_1. This allows us to characterise them immanently:

1   A couple of the form $\langle 0, S(W_1) \rangle$ is always a successor (it succeeds $\langle W_1, W_1 \rangle$). Thus the couple $\langle 0,1 \rangle$ is a successor (it succeeds the minimal couple $\langle 0,0 \rangle$).
2   A couple of the form $\langle 0, L \rangle$ is always a limit: it is the upper bound of the sequence of couples $\langle W_1, W_2 \rangle$ where W_1 and W_2 pass into the limit ordinal L. So that the couple $\langle 0, \omega \rangle$ is the limit of all the couples $\langle m, n \rangle$ where m and n are finite ordinals (and therefore natural whole numbers, see chapter 11).

Couples of the form $\langle W_1, 0 \rangle$ depend just as directly, as regards their intrinsic characterisation, on the nature of the ordinal W_1:

1   A couple of the form $\langle S(W_1), 0 \rangle$ is the smallest couple to have $S(W_1)$ as Max. *in first position*. It is larger than all those which have $S(W_1)$ as Max. *in second position* – that is, couples of the form $\langle W_2, S(W_1) \rangle$ where W_2 is smaller than $S(W_1)$. In fact it comes just after the largest of these couples, which latter will evidently have the largest possible first term to conserve $S(W_1)$'s status of maximum in second position. This largest first term is W_1, the immediate predecessor of $S(W_1)$. The largest of the couples which come before $\langle S(W_1), 0 \rangle$ is therefore the couple $\langle W_1, S(W_1) \rangle$. We can conclude: every couple of the form $\langle S(W_1), 0 \rangle$ is a successor. Thus the couple $\langle 1,0 \rangle$ is a successor (it succeeds $\langle 0,1 \rangle$).
2   A couple of the form $\langle L, 0 \rangle$ is larger than every couple of the form $\langle W_1, L \rangle$ where W_1 is less than L. But there is no such couple that is larger than all the rest, because there is no W_1 that is 'closer' to the limit ordinal L than all others (see **9.18**). The couple $\langle L, 0 \rangle$ is, moreover, smaller than all the couples of the type $\langle L, W_1 \rangle$ where W_1 is not 0. In a sense, it makes a cut between the couples $\langle W_1, L \rangle$ and the couples $\langle L, 0 \rangle$. All the same, amongst the latter there is a minimal couple, which is the couple $\langle L, 1 \rangle$, and which therefore succeeds $\langle L, 0 \rangle$. Here again we find the striking dissymmetry, characteristic of the ordinals, between minimality (guaranteed) and maximality (which presupposes succession). The couple $\langle L, 0 \rangle$ is the limit, or upper bound, of the sequence $\langle W_1, L \rangle$ for $W_1 \in L$, and it immediately precedes the couple $\langle L, 1 \rangle$. It creates an infinite adherence to its left, or 'on this side' of it, and the void of one single additional step to its right, beyond it.

17.12. Let's now turn to 'homogenous' couples of the type $\langle s_1, s_2 \rangle$ or $\langle L_1, L_2 \rangle$. Everything will once more depend upon the Max. of these couples:

- If s_1 or L_1 are the Max., the problem is trivial: $\langle s_1, s_2 \rangle$ is a successor. Just a moment of reflection will show that it comes just after the couple constituted by s_1 (the Max.) and the predecessor of s_2. As for $\langle L_1, L_2 \rangle$, it is surely the limit of the sequence of couples of the type $\langle L_1, W_1 \rangle$, where W_1 traverses the elements of the ordinal L_2.
- If s_2 or L_2 are the Max., things are not much more difficult. It is certain that $\langle s_1, s_2 \rangle$ is a successor: it comes just after the couple constituted by the predecessor of s_1 and by the Max. s_2. As for $\langle L_1, L_2 \rangle$, it is assuredly the limit of the sequence of couples $\langle W_1, L_2 \rangle$, where W_1 traverses the elements of L_1, from 0 'up to' L_1.

17.13. We will finish with mixed couples. The method does not change at all:

- If, in a couple of the type $\langle s, L \rangle$ or $\langle L, s \rangle$, it is L which is the Max., these couples are successors: they come just after the couples obtained by replacing s with its predecessor.
- If s is the Max., the couples are limits of the sequences of couples of the type $\langle s, W_1 \rangle$ or $\langle W_1, s \rangle$, where W_1 traverses the elements of the limit ordinal.

17.14. Finally, we now have a table of immanent characterisations of couples as follows:

Type	Max.	Example	Character
$\langle 0,0 \rangle$	0	Unique	Special
$\langle 0,s \rangle$	S	$\langle 0,1 \rangle$	Successor
$\langle 0,L \rangle$	L	$\langle 0,\omega \rangle$	Limit
$\langle s,0 \rangle$	s	$\langle 1,0 \rangle$	Successor
$\langle L,0 \rangle$	L	$\langle \omega,0 \rangle$	Limit
$\langle s_1,s_2 \rangle$	s_1	$\langle 2,1 \rangle$	Successor
$\langle s_1 s_2 \rangle$	s_2	$\langle 1,2 \rangle$	Successor
$\langle L_1,L_2 \rangle$	L_1	$\langle \omega_1,\omega \rangle$	Limit
$\langle L_1,L_2 \rangle$	L_2	$\langle \omega,\omega_1 \rangle$	Limit
$\langle s,L \rangle$	s	$\langle S(\omega),\omega \rangle$	Limit
$\langle s,L \rangle$	L	$\langle 1,\omega \rangle$	Successor
$\langle L,s \rangle$	s	$\langle \omega,S(\omega) \rangle$	Limit
$\langle L,s \rangle$	L	$\langle \omega,1 \rangle$	Successor

This table has a perfect symmetry, broken only by the inaugural couple of the void with itself, the ontological basis of the whole construction.

17.15. It is entertaining to visualise the beginning of the sequence of ordinal couples.

We have already seen that after the couple $\langle 0,0 \rangle$ comes the couple $\langle 0,1 \rangle$, then the couple $\langle 1,0 \rangle$. One can quickly see that it is $\langle 1,1 \rangle$ that succeeds $\langle 1,0 \rangle$, since it is the largest couple whose Max. is 1. Coming next is $\langle 0,2 \rangle$, which is, as we have remarked, the smallest couple whose Max. is 2. The readers can exercise themselves by calculating the rest. If we draw the succession of couples onto a squared background, using the horizontal axis to represent the ordinal that occupies the first place and the vertical to represent that which occupies the second, we obtain the following:

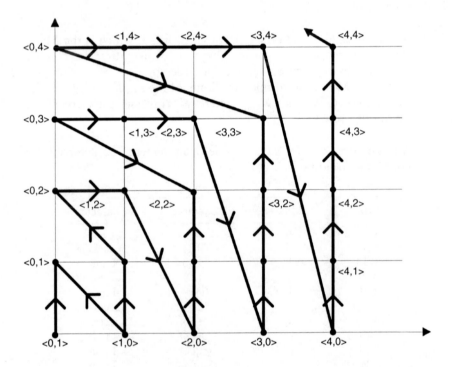

What we see in this diagram is that the route through the couples forms a kind of 'chain' which evidently could be projected onto an ordinal axis. At any given moment we know how to 'produce' the nth couple, as soon as its predecessor has been determined. It is tempting to formalise this intuition by establishing a term-by-term correspondence between ordinals and couples of ordinals, since we have seen that the 'passage to the limit' represents no obstacle to our doing so: there is a concept of the limit couple, structurally distinct from the successor couple. This magnificent construction, which projects couples (representable on a plane or a surface) onto the linearity of their constituents (single ordinals), is a triumph of ontology. It shows that there is no more in the double than in the single. It linearises the divergence of twoness.

17.16. Our underlying motive here is to show that couples of ordinals behave 'like' ordinals themselves. The simplest way is to establish between couples of ordinals and ordinals a biunivocal correspondence (see **4.5**). However, it is dubious, absurd even, to speak of a correspondence or a function between two inconsistent multiplicities. Neither the ordinals nor the ordinal couples are sets. How can we justify comparing or linking these two untotalisable collections?

We have given the principle for the forcing of this impasse in chapter 10: we must, if we can, define the correspondence between the ordinals and the couples via transfinite induction, or recurrence. The function will only be defined at successive levels, without us having to consider the 'alls' between which it operates.

17.17. Let $f(\langle W_1, W_2 \rangle)$ be the function we wish to define and which, to every couple of ordinals, will make correspond biunivocally an ordinal: $f(\langle W_1, W_2 \rangle) = W_3$.

We are firstly going to root the function f securely in its first value, which will correspond to the smallest of the couples, the couple $\langle 0,0 \rangle$. Refer back to chapter 10 for the whole of this procedure.

We posit explicitly:

RULE 1 $f(\langle 0,0 \rangle) = 0$.

We will then examine the case of successor couples (compare the typology of couples in **17.14**). Let C_2 be a couple which succeeds couple C_1, which we will denote – in an extension of the notation adopted for the ordinals – by $C_2 = S(C_1)$. The simplest way is to make correspond to the couple, via f, a C_2, which is the successor of couple C_1, the successor ordinal of the ordinal which corresponds, via f, to C_1: we make the ordinals succeed 'in parallel' to the succession of couples. We thereby respect the basic idea of induction, or recurrence: supposing the function f to be defined for the couple C_1, we define it by an explicit rule for the couple C_2 which succeeds C_1. We therefore posit:

RULE 2 $f(C_2) = f(S(C_1)) = S(f(C_1))$.

Let's move on to limit couples. We suppose the function f to be defined for all couples that precede a limit couple CL. To all these couples, the function f makes correspond an ordinal $W = f(C)$. The idea is evidently to take, as value of f for the limit couple CL, the ordinal that comes just 'after' all of the ordinals thus associated, via f, with the couples that precede CL. We know of the *existence* of this ordinal that comes just 'after' a set of ordinals (see N6): it is the upper bound of that set, denoted by sup. We posit then that $f(CL)$ is the sup. of all the ordinals $f(C)$ for the set of C smaller than CL. So:

RULE 3 $f(CL) = \text{sup.}(f(C))$, for $C < CL$.

The inductive definition of f is now complete, since we have covered the three cases – the minimum ($\langle 0,0 \rangle$), successors and limits – defining f via an explicit rule which makes its value depend upon the values obtained 'below' the term in question.

17.18. A few examples.

What, for example, is the value of $f(\langle 0,1 \rangle)$? We have seen that the couple $\langle 0,1 \rangle$ is the successor, in the order of couples, of the couple $\langle 0,0 \rangle$. We apply rule 2: $f(\langle 0,1 \rangle) = S(f(\langle 0,0 \rangle))$. But rule 1 indicates that $f(\langle 0,0 \rangle) = 0$. Then it must be: $f(\langle 0,1 \rangle) = S(0) = 1$.

What is the value of $f(\langle 0,\omega \rangle)$? We have seen that the couple $\langle 0,\omega \rangle$ comes just after the set of all the couples $\langle m,n \rangle$, where m and n are finite ordinals (the natural whole numbers). Now it is clear that f associates a finite ordinal with each of these couples, since a successor couple will be associated with the successor of the ordinal that corresponds to its predecessor, and since one begins from 0. To couples of the type $\langle m,n \rangle$ will correspond the sequence 0,1,2, etc. Consequently, $f(\langle 0,\omega \rangle)$ will have as its value the upper bound of all the finite ordinals, that is, the first infinite (or limit) ordinal, which is to say ω. Thus $f(\langle 0,\omega \rangle) = \omega$.

These elementary examples demonstrate that we are indeed in a position to calculate f for any couple whatsoever: it is enough to 'progress' the length of the well-orderedness of couples. The value for the first couple being fixed, rules 2 and 3 allow us to know the value of f for a couple C on the basis of the values which f assigns to the couples which precede it.

17.19. That our function f, defined inductively with these three rules, is definitely biunivocal merits verification, whatever evidence we may already have on this point.

It must first of all be confirmed that f is injective, or, in Dedekind's terms, distinct (see **4.5**). In other words that, if couple C_1 is different from couple C_2, then ordinal $f(C_1)$ is different from ordinal $f(C_2)$. We can assure ourselves of this by casting our eye over the rules of induction. If two couples are different, they are ordered; say, $C_1 < C_2$. The value of $f(C_2)$ depends on the value of f for the couples which precede it, and it is different from all of these values. Specifically, it is different from the value of f for C_1, which comes before C_2. We can therefore be sure that $C_1 \neq C_2 \rightarrow f(C_1) \neq f(C_2)$. Function f is injective.

In fact, we have a stronger property here: the function 'projects' the order of couples into the order of ordinals (technically, it is a homomorphism from the order of couples into the order of ordinals), such that, if $C_1 < C_2$, then $f(C_1) \in f(C_2)$. For, if C_2 comes after C_1, its value for f (which is either the successor of the value of the couple which precedes it, or the upper bound of the values of f for all the couples which precede it) in any case surpasses the value of f for C_1. Consequently: $C_1 < C_2 \rightarrow f(C_1) \in f(C_2)$.

It remains to be shown that function f is *surjective*, a modern word meaning that every possible value of the function is effectively fulfilled. In other words that, for *every* ordinal W, there exists an ordinal couple C for which $f(C) = W$.

Suppose that an ordinal W exists whose value for function f is not a couple C. Then there exists a smaller such ordinal (principle of minimality), say w. Thus all ordinals smaller than w do correspond, via, f, to a couple. We can see then that w must necessarily also, contrary to the hypothesis, correspond, via f, to a couple. Because, if w is a successor, which means that $w = S(w_1)$ and $f(C) = w_1$, it must then be (rule 2 of the inductive definition of f) that $f(S(C)) = S(w_1) = w$. And, if w is a limit, then, since all the ordinals which precede w correspond, via f, to couples, w itself appears as the upper bound of all those ordinals, and thus, from rule 3, its value for f will be the couple that comes 'after' all the couples corresponding to ordinals smaller than w.

So, finally, f is indeed a biunivocal correspondence between couples and ordinals. This correspondence is, in addition, an isomorphism between the structure of order of couples (via the Max., the first term, and then the second term) and the structure of order of the ordinals (belonging). Suffice to say at this point that the ordinal couples are a sort of 'doubled' image of single ordinals. Taken 'two by two', nature is still similar to itself. Nature is its own mirror.

17.20. These wanderings in nature through the looking-glass of the double teach us:

– that there exists a well-orderedness over couples of ordinals, such that these couples obey, as do ordinals, the principle of minimality;
– that we can speak, as for the ordinals, of successor couples and limit couples, and that these attributes can be uncovered by immanent examination alone of the structure of the couples which possess them;
– that there exists between couples and ordinals a function f which has all the characteristics of a biunivocal correspondence, except that the totalities between which this function operates are inconsistent, so that f must be defined by transfinite induction;
– that this function f defines an isomorphism between the structure of order of couples of ordinals and the structure of order of the ordinals, so that $C_1 < C_2$ implies that $f(C_1) \in f(C_2)$.

In the mirror of the double, nature perseveres in all of its formal comportments.

Identical means would allow us to establish that triplets of ordinals, of the form $\langle W_1, W_2, W_3 \rangle$, have the same properties as couples do, and in particular that they are in biunivocal correspondence with the single ordinals. The same goes for n-tuplets of ordinals of the form $\langle W_1, W_2, \ldots, W_n \rangle$. In matter, it is only the first step that costs. Doubled, nature maintains its order. Reduplicated in finite series as long as you like, nature persists in maintaining its first identity. Stability, homogeneity, order, minimality, the ontological hiatus between successors and limits: all of this remains when the simplicity of the ordinal is multiplied within the limits of the finite. Nature is its own hall of mirrors.

17.21. Mallarmé wrote: 'Nature is there, it will not be added to'.[3] And it is a fact that, if one adds to nature, and even if one adds and adds, and so on repeatedly, the domain of natural multiples attests unabated to the pregnancy of the Same. This is what we grasp in every experience of the natural: that ramified growth, reproductive division, far from suggesting to us the Other, reposes in itself, in the eternal seat of its order.

Now, we know that every operation, every algebra, is concerned with a doubling or tripling of the terms upon which one operates. We add two numbers to obtain a third, calculate the smallest common divisor of two numbers, arrange in a finite sequence the components of a polynomial . . . All these disciplines of reckoning and algebra have as their substructure a finite listing of numerical marks.

If it is true that natural multiples, ordinals, furnish the matter of Number, we can understand why the *possibility* of operations, of algebra, of reckoning, finds its ontological guarantee in nature's capacity to maintain the identical within division. Beneath the apparent variegation of schemes of reckoning, the variety of operations and of algebraic structures, lies this perseverance of natural being, this immanent stability in finite seriality. An operation is never anything more than the mode in which *our* thinking accords with Mallarmé's maxim: if, without exposing ourselves to the disintegration of the Other, we can combine two Numbers – 'add' one to the other – it is because nature, taken as double, added to itself, re-attached to itself, maintains the immanent form of the multiple–beings through which it inconsists.

An operation, a counting, an algebra, are only marks of our thought's being caught in the mirror-games which it pleases being to proffer, under the law of the Same to which natural multiples dispose it.

18

Algebra of Numbers

18.1. We must finally come to counting.

Once its being has been fixed, the combinatory capacity of Number is a mere consequence. It arises from an investigative ingenuity as to the ways in which couples or triplets of Numbers can be linked. But the source of these links is held completely within the concept by means of which Number is anchored in being. All that operations can do is to deploy – in the numberless domain of Number – the prodigality of being in its possible connections.

Concomitantly, the difficulty resides in the choice of 'good' definitions of the links, so that they should conform to the facilities of calculation: we wish the operations to be associative, for there to be a neutral element, inverses, and it would help if they were also commutative. We would be even happier if operations combined well-behavedly amongst themselves, with a distributivity of one with regard to the other. To arrive at these results, Number must be scrutinised and we must carefully authenticate the links we wish to define.

18.2. The substantial results to be obtained through the ingenuity of operational definitions are as follows:

1   We can define a first operation on Numbers named addition and denoted by +, which has the properties of a commutative group:
   –   associativity: $N_1 + (N_2 + N_3) = (N_1 + N_2) + N_3$ (one can count 'in any order', and achieve the same result);

- a neutral element (which is Zero): $N_1 + 0 = N_1$;
- inversion (which is the symmetric counterpart): $N_1 + (-N_1) = 0$;
- commutativity: $N_1 + N_2 = N_2 + N_1$.

2 We can define on Numbers a second operation, named multiplication, and denoted by $\cdot$, which has the following properties:
- associativity: $N_1 \cdot (N_2 \cdot N_3) = (N_1 \cdot N_2) \cdot N_3$;
- neutral element (which is the Number 1): $N_1 \cdot 1 = N_1$;
- existence of an inverse i(N) for every Number different from Zero: $N_1 \cdot i(N_1) = 1$;
- commutativity: $N_1 \cdot N_2 = N_2 \cdot N_1$.

3 Multiplication is distributive in relation to addition: $N_1 \cdot (N_2 + N_3) = (N_1 \cdot N_2) + (N_1 \cdot N_3)$.

These three operational considerations would lead us to say that Numbers *form a commutative field*, if it were not for one problem: Numbers *do not even form a set*, because they are inconsistent. How can something be *a* field – which is supposed to be an algebraically defined entity – if it cannot be counted as *a* multiple?

Therefore, prudently, we will say only this: that every set constituted of Numbers whose matter is less than a given cardinal infinity (therefore every set constituted from Numbers whose matter is bounded by a 'brute' fixed infinite quantity) can be given the structure of a commutative field.[1] What is more, as can be proved for the rationals or the reals, there are other sets of Numbers that are also commutative fields. Numerical inconsistency can be 'sectioned' into innumerable algebraic structures.

These logical caveats aside, the algebra of Numbers is the richest conceivable: its calculative capacities equal – for example – those with which the real Numbers furnish us (in particular, it can be proved that every Number has a square root, which is not the case if one is operating, for example, in the field of rational numbers).

18.3. A result at once laborious (in its procedures of verification) and of key importance (for the validity of our concept of Number) is the following: operations defined on Numbers coincide with operations defined on 'our' numbers, if the latter are thought in their being as Numbers. In other words: take two real numbers r_1 and r_2, taken in their usual algebraic sense. If the sum of r_1 and r_2, such as we know it, is the real number r_3, then the sum of r_1 and r_2 such as it is represented 'itself' in Numbers (with Numbers of matter finite or equal to ω, see **16.28**) – 'sum' being taken in the sense of the addition defined on Numbers – will be precisely the representative, within Number, of the number r_3. The same will go for multiplication, etc. More technically, we can say that the field of the reals, as we know it in classical analysis, is isomorphic to the reals thought as subset of Numbers.

It is not, therefore, solely in their being that 'our' usual numbers can be thought of as singular types of Numbers, but also in their algebra. Our real Numbers are ultimately *indistinguishable* from real numbers. In particular, real Numbers constitute a complete ordered Archimedean field, which is the univocal customary definition of real numbers.

It can be said, ultimately, that all the dimensions and capacities of 'historical' numbers are retained by their presentative instance in the innumerable swarm of Numbers. Which confirms:

–   that the ontological essence of a number is nothing more than that which our thought apprehends it to be when it is determined as a type of Number;
–   that the operational or algebraic properties are only the effect of a correct determination, on the basis of natural multiplicities, of the *being* of Number.

We therefore find the programme of unification of the concept of Number (one sole concept which subsumes the natural whole numbers, the negative whole numbers, the rationals, the reals and the ordinals) to be wholly realised, firstly in multiple–being, and then in the operational dimensions.

It is now possible for us to speak freely of, and to submit to calculation, entities previously devoid of any sense, like the sum of an ordinal and a real number, or the division of a transfinite ordinal by a rational number, or the square-root of the division by three of an ordinal, etc. Incredible equations like:

$$\frac{\frac{1}{2}\omega_1 + \sqrt{\pi + \omega^2 + 7\left(x^{\omega+5} + 2\omega\right)}}{\left(\omega^7 + 2^\omega\right)\left(\frac{1}{3}S(\omega) - \frac{17}{5}\right)}$$

– which, in the dispersed and lacunary historical theory of numbers, would have made absolutely no sense – in the unified framework of the concept of Number become perfectly meaningful algebraic formulae, indicating certain procedures of calculation and definite results.

Number thus founds in being the literal connection of what, under the disparate name of 'numbers', had defined heterogeneous domains.

18.4. The definition of operations on Numbers is essentially a technical affair. Whoever wishes to follow it in all its detail is referred to the literature.[2] Nevertheless, its animating spirit allows a revision of concepts, a final passage through the Idea of Number. In particular, the systematic use of transfinite induction highlights the fact that Number, thought of in its being, is essentially an infinite multiple (the section of a form from an infinite ordinal–matter). In the same way, the recourse to sub-Numbers of a Number in order to construct operations 'from below' attests to the importance of the fact that every Number can be presented as a cut of its low set and its high set (see chapter 14). And again, it is by presenting the result of an operation as a cut (see chapter 15) – that is, by utilising the fundamental theorem – that we can handle induction. Lastly, the correlation explored in chapter 17 between couples of ordinals and ordinals plays a major role in this whole process – as one might expect (since an operation connects *two* Numbers). So as not to forego these recapitulations in thought, we will cover the essentials of the definition of addition.

18.5. The general idea is as follows: given two Numbers N_1 and N_2, we can make them correspond to two ordinals W_1 and W_2 simply by taking their respective matters, $W_1 = M(N_1)$ and $W_2 = M(N_2)$. We know that a certain ordinal corresponds to these two ordinals via the biunivocal function f, which associates an ordinal with every couple $\langle W_1, W_2 \rangle$ of ordinals (see **17.17**). This ordinal will fix the 'level' of definition of the additive operation: we will suppose that addition is defined for all couples of Numbers N_3 and N_4 of matter W_3 and W_4, such that the ordinal that corresponds via f to the couple $\langle W_3, W_4 \rangle$ is *smaller* than the ordinal associated with the couple $\langle W_1, W_2 \rangle$. We then propose an explicit rule, which will define the sum $N_1 + N_2$ on the basis of sums of the type $N_3 + N_4$, defined at a lower ordinal level.

Now, such sums are given by the sub-Numbers of N_1 and N_2. A sub-Number, being a 'partition' of a Number for an ordinal smaller than its matter, has an inferior matter.

We can then pass on to the next stage, which is the core of the construction. Take the Numbers N_1 and N_2, of matter W_1 and W_2. Consider a sub-Number N_1/w_3 of N_1, and a sub-Number N_2/w_4 of N_2. Now take the couples $\langle w_3, W_2 \rangle$, or $\langle W_1, w_4 \rangle$. I say that they are lower than the couple $\langle W_1, W_2 \rangle$, by the rules of order of couples, and remembering that w_3 and w_4 are respectively smaller than W_1 and W_2 (see **17.6**). This is an excellent exercise for the reader, but see the note.[3]

As the function f is an isomorphism of the order of couples with the order of the ordinals, we will also have:

$$f(\langle w_3, W_2 \rangle) \in f(\langle W_1, W_2 \rangle),$$

and

$$f(\langle W_1, w_4 \rangle) \in f(W_1, W_2).$$

Which is to say that *the ordinal level associated with couples of Numbers of the type* $\langle N_1/w_3, N_2 \rangle$, *or* $\langle N_1, N_2/w_4 \rangle$ *will always be lower than the ordinal level associated with couples of Numbers* $\langle N_1, N_2 \rangle$.

Given this fact, in order inductively to define the sum of N_1 and N_2, we can suppose defined sums of the type $N_1/w_3 + N_2$, or $N_1 + N_2/w_4$, which pertain to a lower ordinal level. We will thus pass on to the definition of $N_1 + N_2$ by formulating a rule which assigns the value of this sum on the basis of the various values between N_1 and N_2 on the one hand, the sub-Numbers of N_1 and N_2 on the other. The immanent concept of sub-Number will serve to underwrite the induction, which fixes their ordinal level on the basis of a couple formed of the matters of the two Numbers under consideration.

Finally, the strategy will mobilise the fundamental theorem of the cut. We will begin with the low set and the high set of the two Numbers N_1 and N_2. We suppose defined the sums of each of the two Numbers with the sub-Numbers of the low set and of the high set of the other Number, according to a fixed combination. These sums can be assumed, because their ordinal level is lower. We can thus obtain two sets of Numbers, and the sum of N_1 and N_2 will be the unique Number defined as cut of these two sets.

18.6. Inductive definition of the addition of two Numbers

'Level zero' of the induction contains only the Number (0,0). It is the only one to have 0 as matter. We can thus posit:

RULE 1 $(0,0) + (0,0) = (0,0).$

We will now suppose that addition is defined for all levels lower than an ordinal W, that is, all levels corresponding to Numbers N_3 and N_4 (taken in that order) such that, their respective matters being W_3 and W_4, it is the case that $f(W_3, W_4) \in W$.

Now take a couple of Numbers N_1 and N_2 such that, their respective matters being W_1 and W_2, it is the case that $f(W_1, W_2) = W$. In other words a couple of Numbers belonging to ordinal level W.

We have remarked that all the couples of type N_1 and N_2/w, or N_1/w and N_2, where N_1/w and N_2/w are sub-Numbers of N_1 and N_2, belong to ordinal levels inferior to those of the couple N_1 and N_2, and therefore inferior to W.

It follows that we can suppose defined all the additions of the type $N_1 + N_2/w$, or $N_1/w + N_2$.

We must agree on an important written convention here. We will write $N_1 + Lo(N_2)$ for *the set of Numbers* constituted by all the results of the addition of N_1 with each of the Numbers of the low set of N_2 (the low set of N_2 is constituted, remember, of all sub-Numbers of N_2 smaller than N_2). If $Lo(N_2)$ is empty, the Number denoted by $N_1 + Lo(N_2)$ would be undefined (we will not consider this in the calculations).

In the same way, we write $N_1 + Hi(N_2)$ for the set of Numbers constituted by all the results of the addition of N_1 with each of the Numbers of the high set of N_2 (the high set of N_2 being constituted by all sub-Numbers of N_2 larger than N_2). The convention will always be not to bother writing this if $Hi(N_2)$ is empty.

We will adopt the same notation to designate sets of Numbers which result from additions implicated in $Lo(N_1) + N_2$, or $Hi(N_1) + N_2$.

Addition will then be defined as follows: on the one hand we take the set of Numbers constituted by all the Numbers of $Lo(N_1) + N_2$, together with all the Numbers of $N_1 + Lo(N_2)$; on the other hand, the set constituted by all the Numbers of $Hi(N_1) + N_2$, together with all the Numbers of $N_1 + Hi(N_2)$. In other words, we 'collect' on one side the Numbers which are the sum of N_1 and N_2 and the low sub-Numbers of the other Number, and on the other side the same sums, but with the high sub-Numbers.

We thus obtain two sets of Numbers, which we can call L and H.

It is not hard to prove, by way of a 'incremental' induction which I leave to one side,[4] that L and H are in a situation of a cut: every Number of L is smaller than every Number of H.

We then utilise the fundamental theorem (chapter 15). The result of the addition of N_1 and N_2 will be precisely the Number which makes a cut between these sets, that is, the unique Number of minimal matter situated between the sets:

$$L = (Lo(N_1) + N_2, N_1 + Lo(N_2))$$
$$H = (Hi(N_1) + N_2, N_1 + Hi(N_2))$$

We posit:

RULE 2 $N_1 + N_2 = $ L/H, cut of the two sets defined above.

18.7. Addition is commutative
In fact, this cut, which supposedly defines the sum $N_2 + N_1$, operates on the same sets as the cut which defines $N_1 + N_2$, as one can show inductively with no difficulty.

It is true at level 0, where there is only the sum, certainly commutative, $0 + 0$.

Suppose that the sums of ordinal levels inferior to $f(W_1,W_2) = W$ are commutative. Then, in particular, the sums $Lo(N_1) + N_2$ or $N_1 + Lo(N_2)$ are commutative. So the set $L = (Lo(N_1) + N_2, N_1 + Lo(N_2))$, which serves to define $N_1 + N_2$, is composed of the same Numbers as the set $L' = (Lo(N_2) + (N_1), N_2 + Lo(N_1))$, which serve to define $N_2 + N_1$. Evidently, the same goes for the set H. And consequently, $N_2 + N_1$, being defined by the same cut as $N_1 + N_2$, is equal to it: addition is commutative.

18.8. The Number 0, which is more precisely the Number (0,0), is the neutral element for addition
It is a question of proving that, for every Number N, $N + 0 = N$. Induction can this time be applied directly to the ordinal–matter of the Numbers.

It is true at level 0, since rule 1 prescribes that $0 + 0 = 0$.

Suppose that this is true for all the Numbers of lower matter than W_1. In other words, for every Number N of matter w such that $w \in W_1$, $N + 0 = N$.

Now take a Number N_1 of matter W_1. Let's examine the sum $N_1 + 0$. The sets L and H of the cut which define the addition are:

$$L = (Lo(N_1) + 0, Lo(0) + N_1)$$
$$H = (Hi(N_1) + 0, Hi(0) + N_1)$$

But the low set and the high set of the Number 0 – that is, (0,0) – are empty (0 has no sub-Numbers). The conventions adopted in **18.6** prohibit us from taking into account the terms $Lo(0) + N_1$ and $Hi(0) + N_1$. So we actually have:

$$L = (Lo(N_1) + 0)$$
$$H = (Hi(N_1) + 0)$$

But $Lo(N_1)$ and $Hi(N_1)$ are composed of sub-Numbers of N_1, and therefore of Numbers of lower matter than W_1. Consequently, the hypothesis of induction applies to all the Numbers of $Lo(N_1)$

or of $Hi(N_1)$: for any such Number, say N_1/w_2, it is the case that $N_1/w_2 + 0 = N_1/w_2$.

We can, with a slight abuse of notation, write this result in the form $Lo(N_1) + 0 = Lo(N_1)$, $Hi(N_1) + 0 = Hi(N_1)$. So that, ultimately, L and H, which define by a cut the sum $N_1 + 0$, are no other than $Lo(N_1)$ and $Hi(N_1)$. But the Number defined by a cut between its low set and its high set is precisely the Number N_1 itself, and so it is indeed the case that $N_1 + 0 = N_1$.

The induction is complete: for a Number N, whatever its matter, 0 is a neutral element for addition.

18.9. Every Number N apart from 0 allows the Number −N as its inverse for addition: N + (−N) = 0

An important point: since −N inverts the form and the residue of N, *the low set of −N is composed of the Numbers −N/w, where N/w is a Number from the high set of N; and the high set of −N is composed of the Numbers −N/w, where N/w is a Number from the low set of N.* A sub-Number N/w is in the low set if w is in the form, and it is in the high set if w is in the residue. These determinations will be inverted in −N. And, since everything that precedes w in N is also inverted (what was in the form is in the residue, and what was in the residue is in the form), in addition to the exchange of the low set and the high set, we will also have an exchange of the signs of positive and negative.

In an abuse of notation, we could therefore write the high set of −N as $-(Lo(N))$, and the low set of −N as $-(Hi(N))$.

The result (see the inductive definition of addition) is that the two sets L and H which define by a cut the sum $N + (−N)$ are the following:

$$L = (Lo(N) + (−N), N + (−(Hi(N))))$$
$$H = (Hi(N) + (−N), N + (−(Lo(N))))$$

So the strategy of the proof consists in proving that all the Numbers of L are negative and all the Numbers of H positive. The result is that 0 is situated between L and H and that, being necessarily of minimal matter in that position, it is 0 that occupies the position of the cut between L and H. Consequently, $N + (−N) = 0$.

LEMMA If the sum $N_1 + N_2$ is positive, if $N_1 + N_2 > 0$, then $-(N_1) < N_2$, and $-(N_2) < N_1$.

The lemma is true at ordinal level 0, because at that level it cannot possibly be the case that $N_1 + N_2 > 0$.

Suppose that it is true up to ordinal level W: for every pair of Numbers N_3 and N_4 such that $f(W_3, W_4) \in W$, the property in question holds. I say that it also must hold for every pair of Numbers N_1 and N_2 such that $f(W_1, W_2) = W$.

The sum $N_1 + N_2$ is defined by the cut L/H. If this cut is positive, it is because set L contains positive Numbers,[5] or else the cut would be negative or null (see the argument on cuts in **15.11**). As for set H, it only contains positive Numbers. Consequently, there are Numbers in $Lo(N_1) + N_2$ or in $N_1 + Lo(N_2)$ that are positive, and all the Numbers of $Hi(N_1) + N_2$ or of $N_1 + Hi(N_2)$ are so.

Take for example $N_1/w + N_2$ as a positive Number of $Lo(N_1) + N_2$. The pair of N_1/w and N_2 is of lower ordinal level than W, and the lemma is therefore supposed to be true of it: since the sum $N_1/w + N_2$ is positive, it is the case that $-(N_2) < N_2/w$, and, since N_1/w is in the low set of N_1, it is a fortiori the case that $-(N_2) < N_1$. In examining the other components of sets L and H, the lemma can be established in all generality.

Now let's come back to the sum $N + (-N)$. Consider the set L which defines it by a cut, so:

$$L = (Lo(N) + (-N), N + (-Hi(N))).$$

Suppose that there are positive Numbers in L. Take for example one such Number $N/w + (-N)$, where N/w is from the low set of N. In virtue of the lemma, it is the case that $-(-N) < N/w$, so $N < N/w$, which is impossible since N/w, being from the low set of N, must be smaller than N. If $N + (-N_1/w)$ is positive, N_1/w being in the high set of N, it must be the case that $N_1/w < N$, which is prohibited, since N_1/w belongs to the high set. We meet with an impasse, and so there are no positive Numbers in set L.

Symmetrical deductions would demonstrate that there are no negative or null Numbers in set H.

Finally, the cut L/H which defines the result of the addition $N + (-N)$ operates between a set L of negative Numbers and a set H of positive Numbers. The Number of minimal matter situated between these two sets is necessarily 0, and so $N + (-N) = 0$.

So we can say that $-(N)$ is the inverse of N for addition.

18.10. Confirming that the addition of Numbers is associative is, as always, a tiresome calculation. It is, it is ... To the extent that we have established that Numbers, endowed – so to speak – with addition defined inductively by the cut:

$$N_1 + N_2 = (Lo(N_1) + N_2, N_1 + Lo(N_2))/(Hi(N_1) + N_2, N_1 + Hi(N_2))$$

would form (were it not for the inconsistency of their 'All' . . .) an ordered commutative group, of which the Number Zero (either (0,0) or 0, it's all the same) is the neutral element.

To confirm that the 'representatives' in our Numbers of whole positives and negatives, rationals, reals, ordinals, are in fact these numbers themselves, but thought in their being, we must prove that addition (in the normal sense) of these numbers coincides with addition of their being as Numbers. For example, if r_1 and r_2 are numbers from the real field, and if $r_1 + r_2 = r_3$ with 'classic' addition, then the Numbers r_1, r_2 and r_3 defined as Numbers of finite matter or of matter ω, as we presented them in **16.27**, are such that, addition being defined inductively as above, it is always the case that $r_1 + r_2 = r_3$.

These confirmations of algebraic isomorphy demand no little ingenuity, above all when it comes to multiplication (which labyrinth we will avoid entering into).

18.11. I will content myself with carrying out the verification for natural whole numbers.

Remember (from **16.13**) that a natural whole number n presented as Number is of the form (n,n), where n is a finite ordinal. Recall also (ibid.) that the low set of n is constituted by all the whole Numbers lower than n, and that its high set is empty.

Take two natural whole Numbers (n_1,n_1) and (n_2,n_2). Their sum is formally defined by the cut:

$$(Lo(n_1)) + n_2, n_1 + Lo(n_2)/(Hi(n_1) + n_2, n_1 + Hi(n_2))$$

But, as Hi(n_1) and Hi(n_2) are empty, the sums of set H of the cut are not defined (convention on the definition of addition, see **18.6**). Set H is therefore empty, which amounts to saying that the sum is simply *the upper bound of set L*.

Since Lo(n_1) is the set of Numbers lower than n_1, the sum Lo(n_1) + n_2 is constituted by all the sums $0 + n_2$, $1 + n_2$, ... $(n_1 - 1) + n_2$.

And, in just the same way, $n_1 + $ Lo(n_2) is constituted by all the sums $n_1 + 0$, $n_1 + 1$, ..., $n_1 + (n_2 - 1)$.

The largest Number of these sums is in all evidence the Number $n_1 + n_2 - 1$.

Reasoning by induction: suppose that, up to the ordinal rank which corresponds to the pair of Numbers n_1,n_2 (so, in reality, the Numbers (n_1,n_1) and (n_2,n_2)), therefore up to the ordinal $w = f(n_1,n_2)$, it is true that the sum of wholes as Numbers will be the Number which corresponds to the (normal) sum of the wholes. In particular, that it is true for the pair $n_1,(n_2 - 1)$, which is evidently of a lower ordinal rank than the pair n_1,n_2. It is therefore to be supposed that $(n_1,n_1) + ((n_2 - 1),(n_2 - 1)) =$ the Number which corresponds to the ordinary addition of the numbers n_1 and $(n_2 - 1)$, that is, the Number $(n_1 + (n_2 - 1),n_1 + (n_2 - 1))$, where the sign + denotes the ordinary addition of whole numbers.[6]

Now we come to see that the largest Number in the set L which defines $n_1 + n_2$ is precisely $n_1 + (n_2 - 1)$. In virtue of the hypothesis of induction, this Number is the Number which corresponds to its being written as an ordinary addition – the Number which inscribes the whole number $n_1 + n_2 - 1$.

Now, $n_1 + n_2$ (in the sense of the addition of Numbers) is the upper bound of L. And every upper bound is a Number of the type (W,W), as we have demonstrated in 15.9. The upper bound of L will therefore be the smallest Number of the type (W,W) to be superior to the largest Number in L, which is the Number $(n_1 + n_2 - 1,n_1 + n_2 - 1)$ (where the signs + and – have their traditional meaning, as when dealing with numbers). This Number is evidently $(n_1 + n_2,n_1 + n_2)$, because $n_1 + n_2$ is the finite ordinal which comes immediately after the finite ordinal $n_1 + n_2 - 1$.

Consequently, the sum (in the Number sense) of the two whole Numbers n_1 and n_2 is the Number that represents the number sum $n_1 + n_2$ (in the number sense). The addition of whole Numbers is isomorphic to the traditional addition of whole numbers.

The treatment of whole negative numbers poses no great problem (an interesting exercise). Thus it is confirmed that the whole positive and negative Numbers form a commutative group isomorphic to the additive group of the ring $\mathbb{Z}$ of the algebraic whole numbers.

The reader will have grasped the essence of operational procedures: find a 'good' inductive definition of the links, prove the classic algebraic properties (associativity, commutativity, neutral element, inverse, distributivity . . .), confirm that what one obtains is isomorphic, for the classical numbers represented in Numbers, to the structures which these numbers are endowed with.

However laborious these efforts might be, they lead to the desired conclusion: all the classic algebraic structures (the ring $\mathbb{Z}$ of algebraic whole numbers, the field $\mathbb{Q}$ of rationals, the field $\mathbb{R}$ of reals), and all the 'inconsistent' algebras (addition and multiplication of ordinals) are isomorphic to the substructures discernible within Numbers.

And so it is that all types of numbers, without exception and in their every dimension, are subsumed by the unique concept of Number.

Conclusion

19

In Conclusion:
From Number to Trans-Being

19.1. Number is neither a trait of the concept, nor an operational fiction; neither an empirical given, nor a constitutive or transcendental category; neither a syntax, nor a language game, not even an abstraction from our idea of order. Number is *a form of Being*. More precisely, the numbers that we manipulate are only a tiny deduction from the infinite profusion of Being in Numbers.

Essentially, a Number is a fragment sectioned from a natural multiplicity; a multiplicity thought, as ordinal, in its being qua being.

The linear order of Numbers, like their algebra, is *our* way of traversing or investigating their being. This way is laborious and limited. It exhibits Number in a tight network of links, whose three principal categories are succession, limit and operations. This is where the illusion arises of a structural or combinatory being of Number. But, in reality, the structures are consequences, for our finite thought, of that which is legible in Number as pure multiplicity. They *depose* Number in a bound presentation which makes us believe that we manipulate it like an object. But Number is not an object. Before every bound presentation, and in the un-bound eternity of its being, Number is available to thought as a formal section of the multiple.

We might also say that between Number, which inscribes its section in the unrepresentable inconsistency of natural multiples, and number, which we manipulate according to structural links, passes the difference between Being and beings. Number is the place of the being qua being, *for* the manipulable numericality of numbers. Number ek-sists in number as the latency of its being.

19.2. This only makes it more remarkable, then, that we can have some access to Number as such, even if this access still indicates an excess: that of being over knowledges, an excess manifest in the numberless extent of Numbers, compared to what we can know by structuring the presentation of types of numbers. That mathematics allows us at least to *designate* this excess, to accede to it, confirms the ontological vocation of that discipline. The history of mathematics, for the concept of Number as for every other concept, is precisely the history, interminable in principle, of the relation between the inconsistency of being as such, and what our thought can make consistent of this inconsistency. Mathematics establishes ontology as the historical situation of being. It progresses constantly *within* ontico-ontological difference, bringing to light, as the latency of the structures presented in the ontological situation, an excessive horizon of inconsistency, of which structures are only effects for a finite thought. It is this trajectory which we have reconstructed at one of its points: that which designates, beyond numbers, the inconsistent multiple—eternity of Numbers.

19.3. Number is thus rendered over to being, and subtracted from the humanity of operations or figures of order, which nevertheless it continues to subtend in thought. The task concerning Number, and numbers, can only be to pursue the deployment of their concept within ontico-ontological difference. Number falls within the exclusive purview of mathematics, at least so far as the thinking of number is concerned. Our philosophical project prescribes this exclusivity, and designates *where* Number is given as the resource of being within the limits of a situation, the ontological or mathematical situation.

We must abandon the path of the thinking of Number followed by Frege or Peano, to say nothing of Russell or Wittgenstein. We must even radicalise, overflow, think up to the point of dissolution, Dedekind's or Cantor's enterprise. There exists no deduction of Number, it is solely a question of a fidelity to that which, in its inconsistent excess, is traced as historical consistency in the interminable movement of mathematical refoundations.

The modern instance of this movement attests to the void and the infinite as materials for the thinking of Number. Nevertheless, none of these concepts can be inferred from experience, nor do they propose themselves to any intuition, or submit to any deduction, even a transcendental one. None of them amounts to the form of an object, or of objectivity. These concepts arise from a *decision*, whose written form is the axiom; a decision that reveals the opening of a new epoch for the thought of being qua being. Being asks nothing more of us, at this point, than that we doggedly pursue the inscription – within a revised ontological situation – of that which, in tracing the inconsistent latency of being, faithfully prepares the rupture at a point of that place where it consists *for us*.

19.4. It is then possible to maintain that the contemporary 'banalisation' of number is outside all thought. The reign of number, the portents of which I discussed at the beginning of this book, is intransitive to the mathematical thought of Number. It imposes the fallacious idea of a bond between numericality and value, or truth. But Number, which is an instance of being as such, can support no value, and has no truth other than that which is given to it in mathematical thought, effectuating its historical presentation for us.

If the reign of number – in opinion polls or votes, in national accounts or in private enterprise, in the monetary economy, in the asubjectivising evaluation of subjects – cannot be authorised by Number or by the thinking of Number, it is because it follows from the simple law of the situation, which is the law of Capital. This law assures, as does every law, the count-for-one of that which is presented in the situation, it makes our historical situation consist, but it cannot make any claim to truth: neither to a truth of Number, nor to a truth which would underlie that which Number designates as form of being.

In our situation, that of Capital, the reign of number is thus the reign of the unthought slavery of numericality itself. Number, which, so it is claimed, underlies everything of value, is in actual fact a proscription against any thinking of number itself. Number operates as that obscure point where the situation concentrates its law; obscure through its being at once sovereign and subtracted from all thought, and even from every investigation that orients itself towards some truth.

The result is that all thought necessarily deploys itself today in a retreat with regard to the reign of number, including every thought that tries to make a truth of Number. It is in this sense that we must hearken to Mallarmé's slogan, more pertinent than ever: that of *restrained action*.[1]

This whole meditation on the concept of Number, because it restores it to being, necessitates the inversion of the contemporary judgement such as it is presented under the banner of number. We must say, against this judgement, that *nothing* made into number is of value. Or that everything that traces, in a situation, the passage of a truth shall be signalled by its indifference to numericality. Not so that this indifference can in its turn be made into a criteria, because many projects, having no number, have no truth either. But this indifference is a necessary subjectivity.

The reverse side of the abundance of capital is the rarity of truth, in every order where truth can be attested to: science, art, politics and love.

19.5. But if the true is, on principle, in our situation, subtracted from the reign of number, which is only a law of this situation, what is the origin of this process?

A truth can depend neither upon being as such (this is why it does not signal itself through a Number) nor upon the contemporary situation, which is that of Capital (this is why it does not signal itself through numbers). Its origin is evental. But the event is not non-being, however much it exceeds the resources of situation–being. The best way to say it would be that the event is of the order of trans-being: at once 'held' within the principle of being (an event, like everything that is, is a multiple) and in rupture with this principle (the event does not fall under the law of the count of the situation, so that, not being counted, it does not consist). Evental trans-being is at once multiple and 'beyond' the One – or, as I have chosen to call it, ultra-One. The possibility that there can be a truth, in a situation whose state has wholly succumbed to numbers, depends upon a fidelity, subtracted from numbers, to this ultra-One.

To think Number, as we have tried to do, restores us, either through mathematics, which is the history of eternity, or through some faithful and restrained scrutiny of *what is happening*, to a supernumerary hazard from which a truth originates, always heterogenous to Capital and therefore to the slavery of the numerical. It is a question, at once, of delivering Number from the tyranny of numbers, and of releasing some truths from it. In any case, *restrained action* is the principle of a remote disorder: it establishes mathematically that order is but the all-too-human precarity of a thinking of the being of Number; it proceeds, effectively and theoretically, to the downfall of numbers, which are the law of the order of our situation:

'Like a god, I put in order neither one nor the other . . .'[2]

Notes

Translator's Preface

1 London: Continuum, 2005.

2 Minneapolis: University of Minnesota Press, 2003.

3 London: Continuum, 2004; in particular, ch. 5, 'The Being of Number', represents an extremely condensed gloss of the present work.

4 See for example ch. 2n 4.

Chapter 0   Number Must Be Thought

1 [Dedekind, R., *Was sind und was sollen die Zahlen* (Braunschweig: F. Vieveg, 1888); English translation *The Nature and Meaning of Numbers*, in Beman, W. W. (ed., trans.), *Essays on the Theory of Numbers* (La Salle, IL: Open Court, 1901; reprinted NY: Dover 1963). Badiou's reference is to the translation by J. Milner, with H. Sinaceur's introduction, *Les Nombres, que sont-ils et à quoi servent-ils?* (Paris: Navarin, 1979). All references given below are to the numbered sections of Dedekind's treatise. – trans.]

2 [*aei o anthropos arithmetizei* – 'man is always counting' – in Dedekind, 'Numbers', Preface to the first edn. – trans.]

3 [Marx and Engels, *Communist Manifesto*, translated by S. Moore, with introduction and notes by G. S. Jones (London: Penguin, 2002), p. 222. – trans.]

Chapter 1 Greek Number and Modern Number

1 Consider, for example, the definition of number in Euclid's *Elements* (Book VII, definition 2): Ἀριθμός ἐστιν τὸ ἐκ μονάδων συγκείμενον πλῆθος'. We might translate it as follows: 'A number is a multitude composed of unities.' The definition of number is secondary, being dependent upon that of unity. But what does definition 1, that of unity, say? Μονάς ἐστίν, καθ᾽ἥν ἕκαστον τῶν ὄντων ἕν λέγετσι: 'Unity is that by virtue of which each being is said to be one.' We can see immediately what ontological substructure is presupposed by the mathematical definition of number: that the One can be said of a being in so far as it is.

2 [*trait*: see ch. 2n 3. – trans.]

3 [Conway, J. H., *On Numbers and Games* (London Mathematical Society Monographs no. 6, London: Academic Press, 1976). – trans.]

4 [Knuth, D. E., *Surreal Numbers* (Reading, MA: Addison-Wesley, 1974). – trans.]

5 [Gonshor, H., *An Introduction to the Theory of Surreal Numbers* (London Mathematical Society Lecture Note Series, 110, Cambridge: Cambridge University Press, 1986). – trans.]

6 [Since the language of 'whole' and 'natural' numbers is informal and not always applied consistently, it is worthwhile to set out the usage of the present work, along with the formal mathematical equivalents:

   • *whole numbers*: 0, 1, 2, 3 . . . (the non-negative integers, $\mathbb{Z}^*$).

   • *natural whole numbers*: 1, 2, 3 . . . (the positive integers, $\mathbb{Z}+$).

   • *'relative' whole numbers*: . . . –3, –2, –1, 0, 1, 2, 3 . . . (the integers, $\mathbb{Z}$). – trans.]

7 On the dialectic – constitutive of materialist thought – between algebraic and topological orientations, the reader is referred to my *Théorie du Sujet* (Paris: Seuil, 1982), pp. 231–49.

8 [See Bourbaki, N., *Éléments de mathématique, Livre I: Théorie des ensembles* (Paris: Hermann, 1954); English edn *Elements of Mathematics*, Vol I: *Theory of Sets* (Reading, MA: Addison-Wesley, 1968). – trans.]

9 The theme of the cut, in its concept and its technique, is treated in chapter 15 of this book.

10 [See Dedekind, *Numbers*, §73. – trans.]

11 [Frege, Gottlob, *Die Grundlagen der Arithmetik: Eine logisch-matematische Untersuchung über den Begriff der Eahl* (Breslan, 1884); *The Foundations of Arithmetic*, English translation by J. L. Austin (2nd revised edn, Oxford: Blackwell, 1974). References given

below are to the numbered sections of Frege's text. The first German edn appeared in 1884. – trans.]

12   [See Frege, *Foundations*, §28–§29. – trans.]

13   For a particularly brief introduction to the different types of numbers which are used in modern analysis, refer for example to J. Dieudonné's *Éléments d'analyse, I: Fondements de l'analyse moderne* (Paris: Gauthier-Villars, 3rd edn, 1981), chs 1–4.

14   [*unique nombre qui ne peut pas être un autre*: From Mallarmé's 'Un Coup de dés jamais n'abolira le hasard', translated in E. H. Blackmore and A. M. Blackmore, *Collected Poems and Other Verse* (Oxford: Oxford University Press, 2006), pp. 161–81 as 'A dice throw at any time never will abolish chance' (translation modified). – trans.]

15   Natacha Michel proposes the distinction between 'first modernity' and 'second modernity' in *L'Instant persuasif du roman* (Paris: Les Conférences du Perroquet, 1987).

16   Dedekind, *Numbers*, §64.

17   I give a detailed commentary on the Hegelian concept of number – a positive virtue of which is that, according to it, the infinite is the truth of the pure *presence* of the finite – in meditation 15 of *L'Être et l'événement* (Paris: Seuil, 1988), pp. 181–90 [pp. 161–70 in Oliver Feltham's English translation *Being and Event* (London: Continuum, 2005). – trans.]

18   [See Frege, *Foundations*, §84–§86. – trans.]

19   [Dedekind, *Numbers*, §2. – trans.]

20   [Frege, *Foundations*, §74. – trans.]

21   [Dedekind, *Numbers*, §73. – trans.]

22   [*Tout*. – trans.]

23   [Dedekind, *Numbers*, §66. – trans.]

Chapter 2   Frege

1   The key text for Frege's conception of number is *The Foundations of Arithmetic* [on which see above, ch. 1n 11 – trans.]. The fundamental argument, extremely dense, occupies paragraphs 55 to 86 (less than thirty pages in the cited edition). We must salute Claude Imbert's excellent work, in particular her lengthy introduction. [Badiou refers to Imbert's translation *Les Fondements de l'arithmétique* (Paris: Seuil, 1969). – trans.]

2   [*toute pensée émet un coup de dés*: Mallarmé, 'Coup de dés', p. 181. – trans.]

3   [See Frege, *Foundations*, §§ 46–53. Badiou renders Frege's *Eigenschaft* as *trait*: Although Austin has 'property,' I have used 'trait' here, so as not to lose the distinction present in Badiou's text between *trait* and *propriété*. – trans.]

4   [In Austin's English translation, Frege's term *Gleichzahlig* is rendered as 'equal', but see Austin's note at §67 on possible alternatives: I follow both Austin's advice and Badiou's use of *équinuméricité* in employing *equinumerate* as the term which avoids at once imprecision and ugly neologism. – trans.]

5   [Frege, *Foundations*, §68. – trans.]

6   [Ibid., §74. – trans.]

7   [Ibid., §77. – trans.]

8   [Ibid. – trans.]

9   [Ibid., §74. – trans.]

10   [Ibid. – trans.]

11   [Ibid. – trans.]

12   The letter (written in German) in which Russell makes known to Frege the paradox that would take the name of its author is reproduced in English translation in *From Frege to Gödel*, a collection of texts edited by J. van Heijenoort (Cambridge, MA: Harvard University Press, 4th edn, 1981) p. 124. Russell concludes with an informal distinction between 'collection' [or 'set', German *Menge* – trans.] and 'totality': 'From this [the paradox], I conclude that under certain circumstances a definable collection [*Menge*] does not form a totality.'

13   Zermelo develops his set-theoretical axiomatic, including the axiom of separation, which remedies Russell's paradox, in a 1908 text written in German. It can be found in English translation in van Heijenoort's collection, cited in the preceding note. It comes from *Investigations in the Foundations of Set Theory*, and especially its first part, 'Fundamental Definitions and Axioms', pp. 201–6.

14   [Frege, *Foundations*, §58. – trans.]

15   The subordination of the existential quantifier to the universal quantifier means that, given a property P, if every possible x possesses this property then there exists an x which possesses it. In the predicate calculus: $\forall x(P(x)) \rightarrow \exists x(P(x))$. The classical rules and axioms of predicate calculus permit one to deduce this implication. Cf. for example E. Mendelson's manual *Introduction to Mathematical Logic* (NY: Van Nostrand, 1964), pp. 70–1.

16   [τὸ γὰρ αὐτὸ νοεῖν ἐστίν τε καὶ εἶναι, from Parmenides' poem. – trans.]

Chapter 3   Additional Note on
a Contemporary Usage of Frege

1 Miller's text appears in *Cahiers pour l'analyse*, no 1 (Paris: Seuil, February 1966), pp. 37–49 [translated by Jacqueline Rose as 'Suture (Elements of the Logic of the Signifier)' in *Screen*, 18: 4 (1978), pp. 24–34. – trans.]. One ought to read along with it Y. Duroux's article 'Psychologie et logique' appearing in the same issue (pp. 31–6), which examines in detail the successor function in Frege.

2 Cf. A. Badiou, 'Marque et manque: A propos du Zéro', in *Cahiers pour l'analyse*, no 10 (Paris: Seuil, 1969), pp. 150–73.

3 [*J'y suis, j'y suis toujours.* From Rimbaud's 1872 poem '*Qu'est-ce pour nous, mon cœur, que les nappes de sang*' [translated in *Collected Poems*, ed. and trans. Oliver Bernard (London: Penguin, 1986), pp. 202–3. – trans.].

4 [Miller, 'Suture', p. 40. – trans.]

5 [*méconnue.* – trans.]

6 [Miller, 'Suture', p. 40. Translation modified. – trans.]

7 [Ibid. – trans.]

8 [See Frege, *Foundations*, §§ 26–27. – trans.]

9 [Miller, 'Suture', p. 44. – trans.]

10 [Ibid. p. 46. – trans.]

11 [Ibid. p. 47. Translation modified. – trans.]

12 [Ibid. p. 43. – trans.]

13 [*l'instance de la lettre.* – trans.]

14 [Ibid. p. 44. – trans.]

15 On the typology of orientations in thought, cf. Meditation 27 of *L'Être et l'événement*, pp. 311–15 [pp. 281–5 in the English translation. – trans.].

16 [Miller, 'Suture', p. 40. Translation modified. – trans.]

17 [Ibid. p. 41. – trans.]

18 [Ibid. p. 47. Translation modified. – trans.]

19 ['Matrice', in *Ornicar?* 4 (1975); translated by Daniel G. Collins in *Lacanian Ink* 12 (Fall, 1997): pp. 45–51. – trans.]

20 [Miller, 'Suture,' p. 39. – trans.]

21 [*fourmillement*: if not for its unfamiliarity, the more direct etymological equivalent of the psychiatric term *formication*, designating a prickling or tingling as of ants crawling over the

skin, might carry less of an inappropriate sense of dynamic self-organisation than 'swarming': rather than implying any vital movement, Badiou's *fourmillement* seems to denote 'our' phenomenological registration of the icy 'constellations' of Number. – trans.]

22   For example É. Borel, 'La Philosophie mathématique et l'infini', *Revue du mois*, 14 (1912), pp. 219–27.

Chapter 4   Dedekind

1   The reference text for Dedekind's doctrine of number is *The Nature and Meaning of Numbers* [see ch. 0n 1 above – trans.] The first German edition was published in 1888.

2   [Dedekind, *Numbers*, §1. – trans.]

3   [Ibid., §2. – trans.]

4   [Ibid., §§ 21–25. – trans.]

5   [Ibid., §§ 26–35. – trans.]

6   [Ibid. §71. – trans.]

7   [Ibid. §73. Dedekind's text has ϕ where Badiou uses f – trans.]

8   We might say that Frege is a Leibnizian, Peano a Kantian, and Cantor a Platonician.
   The greatest logician of our times, Kurt Gödel, considered that the three most important philosophers were Plato, Leibniz and Husserl – this last, if one might say so, holding the place of Kant.
   The three great questions posed by mathematics were thus:
   1   the reality of the pure intelligible, the being of that which mathematics thinks (Plato);
   2   the development of a well-formed language, the certitude of inference, the laws of calculation (Leibniz);
   3   the constitution of sense, the universality of statements (Kant, Husserl).

9   [See Dedekind, *Numbers*, Preface to the first Edn. – trans.]

10   [Dedekind, *Numbers*, § 64n. – trans.]

11   [Ibid. – trans.]

12   [Ibid., § 66. Dedekind's text has ϕ where Badiou has f, and a, b rather than s_1, s_2. – trans.]

13   [*ça* – also 'id'. – trans.]

Chapter 5 Peano

1 The reference text for Peano is a text published in Latin in 1889, whose English title is: 'The Principles of Arithmetic'. The English translation of this text is found in J. van Heijenoort (ed.), *From Frege to Gödel*, pp. 83–97.

2 [Ibid., p. 85. – trans.]

3 [Ibid. – trans.]

4 This passage is taken from a letter from Dedekind to Keferstein, dating from 1890. The English translation can be found in van Heijenoort (ed.), *From Frege to Gödel*, pp. 98–103.

5 [Van Heijenoort (ed.), *From Frege to Gödel*, p. 85. – trans.]

6 [Title of Husserl's 1911 'manifesto'; translated in Q. Lauer (ed.), *Phenomenology and the Crisis of Philosophy* (New York: Harper, 1910). – trans.]

7 [Van Heijenoort (ed.), *From Frege to Gödel*, p. 85. – trans.]

8 [Ibid., p. 85. – trans.]

9 [Ibid., p. 94. – trans.]

10 [Ibid. – trans.]

11 [Ibid. (Axiom 6). – trans.]

12 [*froide d'oubli et désuétude, une Constellation*: Mallarmé, 'Coup de dés', p. 181. – trans.]

13 Regarding these questions, one might read the (purely historical) chapter 10 of Robinson, A. *Non-Standard Analysis* (Amsterdam: North-Holland, revised edn 1974). Robinson recognises that 'Skolem's work on non-standard models of Arithmetic was the greatest single factor in the creation of Non-Standard Analysis' (p. 278).

   For a philosophical commentary on these developments, cf. A. Badiou, 'Infinitesimal Subversion', in *Cahiers pour l'analyse*, no 9 (Paris: Seuil, 1968) pp. 118–37.

14 [*ça* – trans.]

Chapter 6 Cantor: 'Well-Orderedness' and the Ordinals

1 Cantor's clearest articulation of his ordinal conception of numbers is found in an 1899 letter to Dedekind. See the English translation of the key passages of this letter in van Heijenoort (ed.), *From Frege to Gödel*, pp. 113–17. Cantor demonstrates an exceptional lucidity as to the philosophically crucial distinction between consistent

multiplicities and inconsistent multiplicities. It is to him, in fact, that we owe this terminology.

2   On this point, you are naturally referred to the work of Alexander Koyré.

Chapter 7   Transitive Multiplicities

1   [*prenez ensemble* – with the intended resonance of *ensemble* ('set'). – trans.]

2   [*découpe*: a 'carving out' or deduction. – trans.]

Chapter 8   Von Neumann Ordinals

1   John von Neumann gave a definition of ordinals independent of the concept of well-orderedness for the first time in a 1923 German article, entitled 'On the introduction of transfinite numbers'. This article is reproduced in English translation in van Heijenoort (ed.), *From Frege to Gödel*, pp. 346–54.

   The definition of ordinals on the basis of transitive sets seems to have been taken up again in an article in English published in 1937 by Raphael M. Robinson, entitled 'The theory of classes, a modification of von Neumann's system' (*Journal of Symbolic Logic*, no 2, pp. 29–36).

2   Throughout this book, the ordinals, denoted in current literature by the Greek letters, will be denoted by the letters W and w, supplemented further on with numerical indices, W_1, or w_3, etc. In general, W or w designate a variable ordinal (any ordinal whatever). In particular, we employ the expression 'for every ordinal W'. The notation with indices is used to designate a particular ordinal, as in the expression 'take ordinal W_1 which is the matter of Number N_1'. The subscripts will be used most often to the left of the sign $\in$, to designate an ordinal which is an element of another, as $w_1 \in$ W (ordinal w_1 is an element of ordinal W).

3   [*la Nature* (as opposed to *nature*). – trans.]

4   The Axiom of Foundation, also called the Axiom of Regularity, was anticipated by Mirimanoff in 1917, and fully clarified by von Neumann in 1925. To begin with, it was a matter, above all, of eliminating what Mirimanoff called 'extraordinary sets', that is, those which are elements of themselves or contain an infinite chain of the type $\ldots \in a_{n+1} \in a_n \in \ldots \in a_2 \in a_1 \in$ E. It was realised only

later that this axiom enabled a hierarchical presentation of the universe of sets.

For a historical and conceptual commentary on this axiom, cf. A Fraenkel, Y. Bar-Hillel and A. Levy, *Foundations of Set Theory* (Amsterdam: North-Holland, 2nd edn 1973), pp. 86–102. For a philosophical commentary, see Meditation 18 of *L'Être et l'événement*, pp. 205–11 [pp. 184–90 in the English translation. – trans.].

5 A good presentation of the fact that belonging ($\in$) orders the ordinals totally (strict order) – in other words that, given two different ordinals W_1 and W_2, either $W_1 \in W_2$ or $W_2 \in W_1$ – can be found in Shoenfield, J. R., *Mathematical Logic* (Reading, MA: Addison-Wesley, 1967), pp. 246–7. This proof is reproduced and commented upon in *L'Être et l'événement* in the third section of Meditation 12, pp. 153–8 [pp. 134–9 in the English translation. – trans.].

6 [*découper*. – trans.]

Chapter 9   Succession and Limit. The Infinite

1 Badiou, A., *Manifeste pour la philosophie*, Paris: Seuil, 1989 [translated by N. Madarasz as *Manifesto for Philosophy* (Albany, NY: State University of New York Press, 1999). – trans.]. The circumstances and the effects of the philosophy's suture to the poem, beginning with Nietzsche and Heidegger, are described briefly in chapter VII, entitled 'The Age of Poets'.

2 [Osip Mandelstam, from his *Tristia* (1922). Badiou quotes Tatiana Roy's French translation: *vers ces prairies infinies où le temps s'arrête*. – trans.]. The instant of Presence is beyond all insistence, all succession. The 'eternal midday' is the trans-temporal limit of time. Here is the conjoint site of the poem and the sacred.

It is not always in this place, it must be said, that Mandelstams's poems establish themselves. For in his most powerful poetry he seeks to think the century, and succeeds in doing so.

Chapter 10   Recurrence, or Induction

1 For the demonstration of the validity of definitions by induction, you are referred to K. J. Devlin's *Fundamentals of Contemporary Set Theory* (NY: Springer-Verlag, 1980), pp. 65–70 ('The principle of recursion').

Chapter 12 The Concept of Number: An Evental Nomination

1 To repeat, the basic text for the study of the numbers called 'surre-als' is Gonshor's *Introduction to the Theory of Surreal Numbers* [see ch. 1n 5 – trans.]. The fact that Gonshor and all current theorists of these numbers, which I call Numbers, see them as a 'macro-field' of the reals results in a presentation quite different from my own.

The inital idea of their creator, Conway, was to define 'surreal' numbers directly by means of cuts. A number will be defined as a pair of two *sets* of numbers, conforming to the conditions of the cut (every number in the set 'to the right' in the pair is smaller than every number in the set 'to the left'). The double circularity of this definition obviously must be questioned (number is defined on the basis of number, and inequality between numbers is mentioned without having been properly defined). The operation that serves to undo this circle is obviously transfinite induction, which makes ordinals appear inevitably on the scene. In fact, Conway presents Numbers on the basis of their canonical representation – that is, in my language, their 'structural' character: they are defined *on the basis of their sub-Numbers*. D. E. Knuth's book *Surreal Numbers* (Reading, MA: Addison-Wesley, 1974) gives a 'pedagogical' version of Conway's presentation in the form of a dialogue. It seeks to rec-reate the mentality of a 'researcher' into the matter, but in fact becomes quite convoluted, since in its exposition the employment of the ordinal series is not made explicit. Besides this, it re-establishes, to my mind to the detriment of the real 'genius' of the invention of Numbers, a creationist and progressive logic (first 'creating' zero, then 1 and –1, etc.)

Gonshor starts from a literal 'coding', whereas, in my quest for the concept and its philosophical deployment, I join a set-theoretical lineage. Technically speaking, Gonshor generalises the development in base 2 of the real numbers. A real number can be presented as an infinite denumerable series of signs 1 and 0. Gonshor's idea is to consider such series *of any ordinal length whatsoever*, rather than limiting them to denumerable series. He then begins with two signs + and –, and calls 'surreal number of length W' a series of such signs indexed to the elements of the ordinal W. The index ordinals affected by the sign + correspond to the elements of what I call the form of the Number, and the index ordinals affected by the sign –, to the elements of the residue. The ordinal 'length' corresponds to what I call the matter of the Number.

As an example: the Number which I write (4,(0,3)), whose matter is 4 and whose form contains the elements 0 and 3, is written by Gonshor as follows: + – – +.

Now, of course surreal numbers and Numbers are 'the same thing'. But we might say that Gonshor treats them as inscriptions, or markings, after the manner of Frege and of Peano's arithmetic. The inspiration here is ideographic. Whereas I approach them from the point of view of their multiple–being, in the Cantorian spirit, my inspiration being ontological, or Platonic.

In fact even the technical development ends up being quite different, although the results can always be translated from one version to the other. For example, it is not insignificant that Gonshor, who, with the signs + and –, is unable to denote an occurence of the void, must invoke an 'empty series' of signs, where I would write (0,0). The conceptual advantage of the ontological approach to Number is that it allows one to dispense with all additional literalisation, with every heterogeneous sign, in favour of the two fundamental set-theoretical relations of belonging ∈ and inclusion ⊂. This doubtless explains why for Gonshor the theory of surreal numbers is a sort of specialist technique, whereas for me it is a wholly natural extension of the ontological vocation of set theory to the concept of Number.

2   Gonshor, *Introduction*, p. 43.

3   [*découpe*. – trans.]

Chapter 13   Difference and Order of Numbers

1   Throughout this book, I call a relation (most often one of order) 'total' when two *different* basic terms of the relation are always bound by this relation. Thus I would say that the relation ∈ is total in the ordinals or that the relation ∈ is total in the Numbers.

Sometimes a relation is called 'total' which is also reflexive, binding each term to itself. This is the case, for example, with the relation ≤ (less than *or equal to*) for the natural whole numbers.

Limiting oneself to my definition, which only demands the relation between different terms, and excludes the relation of self with self (an irreflexive relation, then), is more convenient in the case we are dealing with. Where we speak of an order-relation, we mean to say that its axioms are those of *strict* order.

2   For Gonshor, order is easily presented as lexicographical, since surreal numbers are introduced as series of signs + and – (cf. ch. 12n 1).

3   On this point, cf. Miller, J.-C. *Libertes, Lettre, Matiere* (Paris: Conférence du Perroquet, June 1985).

4   [Paul Celan, from *Zeitgehöft* (1976). [Badiou's reference is M. Broda's translation (Paris: Clivages, 1985): *chiquenaude / dans l'âbime, dans les / carnets de gribouillages / le monde se met à bruire, il n'en tient / qu'à toi*. – trans.]

Chapter 14   The Concept of Sub-Number

1   Category theory is an attempt to reformulate all of mathematics within a structural, non-set-theoretical framework whose starting point is 'objects', which are 'types of structures'; and 'arrows', which are transformations, or morphisms, between structures. The concept of substructure can be understood in terms of that of sub-object. A 'sub-object' is in fact an equivalence class for certain arrows. Cf. for example J. L. Bell's book *Toposes and Local Set Theories* (Oxford: Clarendon Press, 1988), in particular the arguments of pp. 49–58.

2   [*sectionne*. – trans.]

3   ['à la matière près': In mathematics, 'à la *x* près' – English equivalent 'up to *x*' – indicates that abstraction is to be made from a certain class of objects, which for the purposes of a particular statement or definition are to be regarded as a single entity. Thus a certain proposition can be said to be true, or a property to be satisfied, 'up to isomorphism', 'up to rotation', 'up to translation', and so on. In the present context, the 'cut' between the high and low sets uniquely defines a Number, so long as we regard all possible configurations of 'matter' as being subsumed under the aspect of the unique minimal case. In other words, the definition of the cut must be supplemented by the principle of minimality. – trans.]

4   [*encadrement*: an interval in the mathematical sense, as in 'interval around a real number'. – trans.]

Chapter 15   Cuts: The Fundamental Theorem

1   The problem of the cardinality of the set of parts of an infinite set is a central problem for set theory after Cantor. The 'minimal' hypothesis, which says that this cardinality is *the smallest* cardinal larger than that of the initial set – the cardinal successor of the one which measures the quantity of that set – is the famous 'continuum hypothesis', denoted by CH in the English literature on the subject.

Following P. H. Cohen's work, we know that the continuum hypothesis is undecidable on the basis of the classical axioms of the theory. It can be affirmed or denied without any contradiction being introduced.

A particularly lucid text on this problem is K. Gödel's 'What is Cantor's continuum problem?'. The English text has often been republished since its first appearance in 1947; for example in P. Benacerraf and H. Putman (eds), *Philosophy of Mathematics* (Cambridge: Cambridge University Press, 2nd edn, 1983) pp. 470–86.

2 [*dans ces parages du vague où toute réalité se dissout*: Mallarmé, 'Coup de dés', p. 181. – trans.]

3 [Ibid. – trans.]

4 The concept of the cut, and the way in which it specifies the relation between punctual intervention and the continuum of situations, traverses all the procedures of truth. Its occurrence can be remarked in the politics of rupture (revoutionary politics), in the artistic theme of novelty or of modernity, in the scientific theme of crises and refoundations, or in the amorous figure of separation. Every fidelity is also the process of a cut.

5 Dedekind's fundamental text on the idea of the cut, dating from 1872, is 'Continuity and Irrational Numbers' [translated in Dedekind, *Numbers*, pp. 1–24. – trans.].

6 The exposition in Gonshor, *Introduction*, begins with the demonstration of the fundamental theorem. His style is very different: both because, as I have already mentioned, Gonshor adopts a line which is oriented more towards calculation than towards set theory; and because he is not content with a proof of existence, but intends to determine *exactly* the Number that is cut (what is called a 'constructive' proof). This concern for determination entails the examination of a great many cases.

7 We will see in chapter 16 that the upper bound of a set L, being of the form (W_1, W_1), *is an ordinal*. This is a striking result.

8 The lower bound of a set H is in fact the negative of an ordinal, a Number $-(W)$. Cf. ch. 16.

9 For rule 2, the reasoning is exactly symmetrical to that which validates rule 1. Let us take rule 3: we have $Id.(W, Nb)$, and W is in the form of Nb. I put W in the form of Ni. Am I not risking making it so that Ni becomes thus as large as a Number of A? Take Na to be this supposed Number. W must be the discriminant of Ni and of Na, which is to say that it is also the discriminant of Na and Nb. Now W is in the form of Nb, one must therefore have $Na < Nb$, which is not allowed.

The same approach can be applied for rule 4.

Chapter 16   The Numberless Enchantment of the Place of Number

1 As we have indicated in notes 7 and 8 of the preceding chapter, a very interesting 'topological' characteristic of the positive and negative ordinals is that every set of Numbers has an ordinal as its upper bound, and the negative of an ordinal as its lower bound. This can

be explained easily enough, since every ordinal is the cut of itself and the void, and every negative of an ordinal, a cut of the void and itself.

2   Gonshor, *Introduction*, p. 32.

3   [*encadrement. –* trans.]

4   The principle of the isomorphism of orders – that is, of the fact that, if N is a Number of finite matter and RA(N) = *r*, then $N_1 < N_2 \rightarrow RA(N_1) < RA(N_2)$ – is simple enough (note that < is to the left of the implication in the order-relation in Numbers, to the right of the relation of ordinary order in the rational numbers). The result is that, in the decomposition of N in the form $1 + 1 + \ldots - \frac{1}{2} +$ etc., what is added 'at the end' decreases very rapidly. This is a quite simple, normal algebraic calculation.

5   See Gonshor, *Introduction*, pp. 30–1.

6   One might object at this point that our Numbers do not authorise the representation either of complex numbers or of quaternions, upon which physics relies to a considerable extent.

But are complex numbers and quaternions numbers? I think it can be reasonably maintained that, from the moment we take leave of all 'linearity' when we abandon dimension 1, we are dealing with constructions *based on* Numbers rather than with Numbers *per se.*

Basically, the innermost essence of complex numbers is geometrical, it is the 'complex plane' which delivers the truth of these 'numbers'. Around the complex numbers is organised the profound link between pure algebra (the extension of fields) and the ontological scheme of space as topological concept. I am tempted to call complex numbers *operators*, operators whose function in thought is to articulate algebra and topology. Hence the simultaneously combinatorial (a complex number being a *pair* of real numbers) and geometrical character of these 'numbers'. They are in fact numbers *which do not number*, but suggest schemes of representation and inscription which are already, in effect, something very close to a conceptual 'physics'.

Moreover, it seems to me unreasonable to speak of 'numbers' when it is not even possible, in terms of the operational field considered, to say that one 'number' is larger or smaller than another. In short: a field *of numbers* must in my view be an *ordered* field, which neither complex numbers nor quaternions are.

Finally, I restrict the concept of Number, in so far as it is thought of as a form of being, to that which can be deployed according to the intuition of a line. This is made clear by the decisive part played in the definition of the being of Number by that fundamental 'line of being' constituted by the ordinals.

7   See Robinson, *Non-Standard Analysis* [see ch. 5n 13. – trans.].

Chapter 17   Natural Interlude

1   On the mathematical personality Ramanujan, see the great number-theorist G. H. Hardy's autobiographical *A Mathematician's Apology* [Cambridge: Cambridge University Press, 1940. – trans.].

2   On the set-theoretical reduction of relations and functions to the pure multiple, and for an ontological discussion of this point, see *L'Être et l'événement*, Appendix 2, 'A relation, or a function, is solely a pure multiple', pp. 483–6 [pp. 443–7 in the English translation. – trans.].

3   [*La nature a lieu, on n'y ajoutera pas*: Mallarmé, 'De la musique et des lettres', in H. Mondor and G. Jean-Aubry (eds), *Oeuvres complètes* (Paris: Pléiade, 1945) pp. 642–57. Translated as 'Music and Literature', in B. Cook (trans.), *Mallarmé: Selected Prose Poems, Essays and Letters* (Baltimore: Johns Hopkins Press, 1956), pp. 43–56 (translation modified). – trans.].

Chapter 18   Algebra of Numbers

1   It is equally true that every set of Numbers whose matter is lower than *or equal to* a given infinite cardinal is a commutative field. In this regard, Gonshor is right to say that the study of the field of Numbers of finite matter or equal to ω ('of countable length' [see Gonshor, *Introduction*, p. 103. – trans.]) would be most worthwhile. This field allows real Numbers as a subset, but it also contains infinitesimals and cuts of cuts. It would be possible to develop a wholly original analysis here.

2   Gonshor, *Introduction*, Ch. 3, 'The Basic Operations'.

3   Take two ordinals W_1 and W_2, where $f(\langle W_1, W_2 \rangle) = W$. If W_1 is maximal in the couple, every couple $\langle W_1, w_4 \rangle$ where $w_4 \in W_2$ is smaller than the couple $\langle W_1, W_2 \rangle$ in the order of couples (see **17.6**), because they have the same Max (which is W_1) and the same first term (which is also W_1), but the second term of the couple $\langle W_1, w_4 \rangle$ is smaller than the second term of $\langle W_1, W_2 \rangle$. So, $f(\langle W_1, w_4 \rangle) \in f(\langle W_1, W_2 \rangle)$, since f is an isomorphism of order between couples of ordinals and ordinals.

If it is W_2 that is maximal, the same conclusion follows, since the Max of $\langle W_1, w_4 \rangle$ is lower.

Similar verifications can be made for any such case.

4   The induction in question consists of proving simultaneously:

   • that, if $N_2 < N_3$, then $N_1 + N_2 < N_1 + N_3$ (compatibility of order and additive structure);

- that the Numbers of set L are all smaller than the Numbers of set H.

5 To be really meticulous, we must take into account the case where L does not contain any positive Numbers, but does contain 0. In this case, 0 is the *internal* maximum of L. One can take it as 0 'alone', or identify L with the set (0). The reasoning is then much simplified.

6 The reader might be perturbed by the constant amphibolies of notation (the sign < used in one place for the order of Numbers, elsewhere for that of this or that particular type of number, etc.) In fact, mathematicians (who say in such a case 'that there is no possible equivocity') express through such amphibolies their natural tendency to *identify* purely and simply, and therefore to name identically, relations and operations which are defined with isomorphic structures. How else could Category Theory have arisen, taking as its 'primitives' not multiplicities, but 'morphisms', or arrows, designating 'correspondences' between structural 'objects'?

Chapter 19   In Conclusion: From Number to Trans-Being

1 [*action restreinte*: Mallarmé, 'Action restreinte', in *Oeuvres complètes* (see ch. 17n 2), pp. 369–73.]

2 [*Como um deus, não arrumei nem uma coisa new outra*: from Álvaro de Campos *aka* Fernando Pessoa's 1929 poem 'Reticências'. See F. Pessoa, edited by M. A. D. Galhoz, *Obra Poética* (Rio de Janeiro: Aguilar, 1960). This is a variant of his 'Quasi' (cf. Vol. II of the *Ediçao Crítica* (Imprensa Nacional – Casa da Moeda, 1990), p. 215), where we read *Como um deus, não arrumei nem a verdade nem a vida* ('Like a god, I arranged neither truth nor life'). Badiou quotes A. Guibert's French translation: *Tel un dieu, je n'ai mis de l'ordre ni dans l'un ni dans l'autre.* – trans.]

Index

Lightning Source UK Ltd.
Milton Keynes UK

9 780745 638799